Succeeding in Organic Chemistry

A Systematic Problem-Solving Approach to Mastering Structure, Function and Mechanism

Joseph C. Sloop

authorHOUSE®

AuthorHouse™
1663 Liberty Drive
Bloomington, IN 47403
www.authorhouse.com
Phone: 1-800-839-8640

First published by AuthorHouse 8/26/2010

ISBN: 978-1-4520-1737-2 (sc)
ISBN: 978-1-4520-2101-0 (e)

Library of Congress Control Number: 2010907186

Printed in the United States of America
Bloomington, Indiana

This book is printed on acid-free paper.

Preface

What is it about organic chemistry that strikes such fear into the hearts of budding

scientists, doctors and chemists? Why do students approach this subject with trepidation?

Although these questions are not easily answered, over the years of studying and teaching

organic chemistry I have learned how students think about organic chemistry.

The purpose of this book is three-fold: to explode the misconceptions and

misgivings that are prevalent regarding this vast subject, provide additional insight for

students on a number of concepts essential to mastery of organic chemistry, and explore

alternative learning strategies to assist the beginning organic chemistry student in applying a

specialized problem solving technique which centers on structure, function and a

mechanistic approach. The goal for the audience is to, given the appropriate tools, arrive at

the realization that introductory organic chemistry may not only be mastered, but may be

enjoyed as well.

Acknowledgements

A mentor, friend and colleague for many years, Dr. Carl L. Bumgardner gave significantly of his time, energy and counsel in the preparation of this text. The content and form of this work are due in no small part to his efforts, for which I am very grateful. I also would like to thank Lisa Deal Sloop, who supplied the cover photography and designed the cover art for the text. I also express my gratitude to Dr. Richard Pennington, MAJ Doug Andresen, MAJ Richard Comitz and MAJ William Eckels for their assistance in reviewing the text.

Dedication

This book is dedicated to my wife, Lisa. Her unwavering support while I wrote this text over the last five years will always be appreciated.

Table of Contents

Table of Contents (cont)

Table of Contents (cont)

Chapter 1 – Studying for Success

1.1 Exploding the Myth

A frequent comment heard from beginning organic chemistry students is that "I'll never be able to learn this – there's too much to memorize." At least part of this statement is true – there is too much material to commit to memory. Unfortunately, many students attack organic chemistry this way, that is, they never get past the idea that memorization is the only key to success.

For those who take foreign languages, learning a foreign alphabet, numbers, grammatical rules and certain idiomatic expressions are necessary before being able to speak, read and write in that language with some degree of fluency. Much of this requires memorization. It is the continued use of these components of a language and not the fact that you memorized the rudiments that eventually leads to an understanding of how to put words together in spoken and written forms.

Organic chemistry is similar. It has a unique language students must know. So, some memorization of the lexicon early is critical. Without it, we could never get to the good part of organic chemistry – understanding how and why chemical reactions take place. Once we can speak, read and write of organic chemistry using the accepted terminology, grasping concepts that are based in that common language no longer need be just memorized – they can be understood and applied. It is the continued use of these concepts through a systematic method of working and solving problems in organic chemistry that moves the true student of organic chemistry beyond the memorization mode.

1.2 Being the Student of a Subject

The difference between simply taking a course and being a student of a subject cannot be overstated. Although on the surface these may have similar implications, e.g. grades and movement into more advanced science coursework, the road a student takes to negotiate organic chemistry will be established early. What this means is that someone learning about organic chemistry can either see this fascinating subject as the beginning of a journey to greater understanding or, as is often the case, a means to some end – another block checked on the way to graduation, preparation for post-baccalaureate studies, etc.

The truth is organic chemistry is not like the general chemistry courses that most undergraduates study. Although it is a survey of many topics in this field, it is something more important: it is the gateway and linch pin for other chemical and life science courses a student may study. Without a firm grasp of the concepts in organic chemistry, chemical engineering, biology, biochemistry, toxicology, and other advanced chemistry courses do not make as much sense as they should. Once you understand this external thread of continuity, the reason for learning organic chemistry is obvious.

In addition, an internal thread of continuity runs through organic chemistry. Think of it as a grand, unfinished play – one with many acts and scenes, tied together by the plot. There may be numerous characters, but they propel the play forward within the context of the plot. While organic chemistry contains many subjects, certain fundamental concepts of structure and function, if studied and understood early, will propel you, as a student forward toward better understanding of this subject and enable you to solve the many different types of problems you will encounter.

One last thing. Organic chemistry is, as are all sciences, a snapshot in time. As students, we must realize that new discoveries, investigations and research continually change organic chemistry. The dozens of academic journals devoted to organic chemistry produce upwards of 100,000 pages of new research results annually. This may add to the complexity of it, but it also makes clear the relevance of learning the fundamentals, so that we may apply them to new cases we may learn of later.

1.3 Studying Organic Chemistry

Now that you understand a bit more about why you want to be a student of organic chemistry, what steps can be taken to enable your success? As in every course, success in organic chemistry depends on many factors – but a few bear special discussion. These include:

> ➤ knowing and using all available study resources,
>
> ➤ creating an effective study environment and
>
> ➤ implementing study sessions using a review, read and practice, preview methodology.
>
> ➤ Solving problems using a systematic approach.

Let's look at the first of these, the study resources available for your course.

❖ **Knowing and Using All Available Resources**

○ <u>Frame of mind is key.</u> While institutions choose organic chemistry texts and a variety of other resources in print or electronic media to suit the needs of the chemistry curriculum, you bring the most valuable resource to the table - your mindset. As you prepare to study, your frame of mind will determine, in large part, how well you will retain the subject material when studying. Without a positive attitude, the other reference materials will not be of much use.

○ <u>Learning is your responsibility.</u> The teacher cannot force you to learn, you control it. How effectively you study and learn is influenced by your external surroundings and internal level of commitment – factors you determine. By the time you take organic chemistry, you have probably developed some sort of study habits, good and bad. In the end, you must be the judge of what works for you, but remain open to the idea that you may have to "unlearn" some poor study habits and replace them with better ones.

○ <u>Use your visual resources</u> - text packages, electronic media and modeling kits. The study of organic chemistry is visually oriented. A wide variety of media is available to enhance your learning experience. Only if you explore the options can you decide what will help you the most.

More than twenty undergraduate organic chemistry texts are available and in use today. For the most part, these text packages are more than satisfactory, providing coverage of essential theory, practical reactions and discussions of how key transformations take place. In that the choice of text is made for you, your success begins with getting the most out of it.

Nowadays, many texts offer study guides (solutions manuals), electronic media (CD-ROMs), online study resources and even smart cell phone applications such as flash cards for nomenclature and stereochemistry. Your instructor will be able to point you in the direction of these. Many students never discover the wealth of information that is out there. As a student, you should not only ensure that you have access to these resources, but also explore all available materials that come with your text to get the most benefit from them.

Finally, aside from the text, there is one important item that you must have to truly be able to visualize the three-dimensional structure and nature of molecules – a molecular modeling kit. Many types of kits are available; your instructor will probably recommend that you obtain one for use during your course. Because organic chemistry is a visual science, building models of molecules helps you gain a better understanding of their three-dimensional structure. This will aid you in several ways. First you'll be better able to grasp how molecules appear in nature. Next, if you can build it, you can draw it correctly. Finally, when you begin to learn about reactions between organic molecules, models will assist you in visualizing how organic molecules can react to form products.

o <u>Know your chemistry department and professor.</u> You must know what resources your institution's chemistry department offers. Frequently, college chemistry departments will offer old tests, online homework assignments, problem sets and lecture notes to assist you. Nothing replaces working additional problems. Tutorial services available through your school can clarify difficulties you may have in understanding the course material.

More importantly, know your professor – they have a vested interest in your success. This means understanding what he or she thinks is important for you to know. Usually that is provided for you in a syllabus and in some cases lesson and course objectives. Your instructor may have this information posted online in a course website or via applications like WebCT or Blackboard. In addition, many instructors provide airliner videos to help guide students through difficult concepts.

Where possible, obtain copies of previous years' tests to learn the types of questions they ask. Don't be afraid to ask for assistance from your professor early when you are unsure of the material. Do not view your teacher as an impasse to your progress, but as a bridge to better understanding.

❖ **Create and Use an Effective Study Environment**

Having a sound study plan that considers both your environment and study session is essential for success. Study in surroundings that are free from distractions, yet comfortable enough to enable you to focus. Noises, sights and interruptions can make studying less productive, so try to reduce those to a minimum. While soothing music playing on the Ipod may help some to focus, it's not for everyone. You must determine what works best for you. Last, gather all the materials you need to study before you begin so that you won't have to stop to find them.

Plan to study the material for the day's lesson while you are fresh and can devote sufficient time to the task. A good rule of thumb is that you should spend about 1-1½ hours of quality study time for each hour of lecture. Once your environment is set, you will find your study efforts are maximized if you maintain a high level of internal commitment. Your study of organic chemistry includes how you approach both in-class lecture and out-of-class study.

❖ **Study organic chemistry using the review, read and practice, preview technique.**

o <u>Review</u> - First, go over the last day's lesson by reviewing your lecture notes, the pictures and diagrams from the reading material for that lesson, and the problems you worked that amplify the material. Taking about 10-15 minutes to do this will help solidify important concepts emphasized by your professor during lecture and help you discover the relationship between this and the next lesson.

o <u>Read and Practice</u> - Once you have refreshed yourself, begin studying the upcoming lesson by reviewing the lesson outline while scanning the section of text to identify the major points. A common mistake beginning organic chemistry students often make is to read straight through the text highlighting every single passage. Rather, as you scan, consider diagrams, tables and figures first, as these are often used by authors to illustrate key concepts and trends discussed in the body of the text. Unless you understand them, you miss out on an important part of learning organic chemistry – the ability to recognize analogies that help you grasp concepts beyond the material you are presently covering. If online guides, a CD-ROM, your modeling kit or other supplementary learning aids are available to help you understand the material, this is the time to use them.

After you have gleaned the author's main points in the pictures, graphs and tables, begin critically reading the text material, to clarify those topics that you may not have understood fully, highlighting the important areas. Simultaneously, work example problems as you cover the portion of the reading that concerns them. Like diagrams and tables, authors use these problems to illustrate important points brought out in the text. Solving these problems (which usually contain the solutions) provides a check on your level of understanding. If you find your answer does not agree with the author, re-read the portion of the text that explains the concept. Also work assigned homework problems, as most professors will assign problems they believe best illustrate the lesson objectives while helping students grasp the material. If no problems are assigned, select problems that are related to the reading material for the lesson. Attempt to solve the problems on your own first, using the solutions manual only when required to help if you get stumped, and then as a final check of your progress.

In following this read and practice technique, you will have now studied the lesson three ways, by visually reviewing the graphics, by reading and by working problems. This should take about an hour. If you finish with questions, write them down to discuss during the next lecture period.

o Preview - Finally, take another 10-15 minutes, and preview the next day's lesson, focusing on the graphics. The objective here is to determine the linkage between these lessons – again that internal thread of continuity. The result is exposure to a given lesson's material on three occasions, once during the preview, again as you study that part of the text as the lesson of the day, and finally in a review.

❖ **Solve problems using a systematic approach.**

Much of organic chemistry is about solving problems: drawing structures, nomenclature, predicting the products of a reaction or writing a step-by-step explanation for how molecules react. Regardless of the task, students do better when they have a strategy for solving problems. Ultimately, you must decide on the method that works best. Here are a few suggested steps that use the acronym GFPSC.

o G: Write down the *Given* information, draw the structures or reactions given as accurately as possible.

o F: Identify what you must *Find* and write it down. Make sure you answer the question that is being asked.

o P: Devise a *Plan* to solve the problem. What techniques, rules or guidelines do you need? Identify trends or analogous examples that may apply and help you.

o S: *Solve* the problem using the given information and your plan. Ensure your work is complete.

o C: *Check* your work for accuracy. Review all information used in your solution ensuring the data you used was what was required.

1.4 Optimizing your classroom experience.

Attendance in the classroom is paramount in organic chemistry because what is discussed in class one day will often be used at other times throughout the course – the internal thread of continuity. Sit where you can see and hear everything being presented. It is here where knowing your professor really pays off. Be observant – note what he or she emphasizes. Organic chemistry is a visual subject, so if the professor takes the time to discuss it, write something on a board, overhead or present it in a slide show, it is important enough for you to transcribe in your notes or to get a copy.

If, while in class, you have a question, **ask for clarification** before you depart. Students are often afraid to ask questions, but professors are generally pleased when a student shows enough interest to formulate a question. The beginning of understanding is recognizing what you don't know and resolving it.

Likewise, when the professor asks a question of the class, be bold and attempt to **provide an answer**. Answering questions you know has a calming and reassuring effect that puts your mind into a more effective mode for tackling more difficult questions.

1.5 Working additional problems

Once you have studied the reading materials and prepared questions that require answering, when time permits, work additional problems in the text. These are usually found at the end of the chapter. Check the problems to ensure your answers are accurate.

1.6 Seeking additional help if required

If questions remain after study and lecture on a particular concept or problem, seek additional help early. Students at all levels need a jump-start occasionally, so don't be afraid to ask.

1.7 Test Taking

Although taking your first organic chemistry test can be a major source of consternation, it need not be. If you have followed the guidelines set forth in this chapter, a two to three hour review of the major concepts covered in the chapters being tested (with problem solving) will usually be sufficient to garner a good grade. Having said that, here are a few suggestions to assist you during the actual test itself.

- **Don't forget your tools.** Use a sharpened pencil for your test and ensure you have a good eraser. There is nothing worse than a test submitted with scratch-throughs and cross-outs. It is distressing for the instructor to grade tests that are messy. Bring your model kit. Most organic chemistry instructors will encourage you to use your kit, so have it available.
- **Upon receipt of the test, scan through the entire test.** Before answering any questions, identify areas that you understand well and can answer quickly, as well as problems that will require more thought.
- **Before attempting to solve the problem, read the question.** The most common errors on tests result from the student misinterpreting the question. If you're unsure, ask the instructor for clarification.
- **Answer the questions that you know first.** Your objective is to obtain as many points as you can, so don't get bogged down on difficult questions. Come back to them later.
- **Look for clues in the test to answer the questions that are more difficult.** On any given test, a hint or solution to one problem may appear in another section of the test. For example, many organic tests will contain a series of reactions where you may be asked to provide the reactants, reagents or products. Keep this in mind as you begin to answer synthesis questions, since some of the transformations used earlier may be required to complete the synthesis problems.
- **Review your work.** A second look can prevent you from the careless error, whether it be too many bonds around a carbon atom, a misplaced formal charge or an incorrect structure.

Your education is <u>your</u> responsibility, so take ownership of it and get the most from your experience.

Chapter 2 – Fundamentals

2.1. Introduction

This chapter will take a look at some topics that most chemistry students are introduced to in general chemistry, but sometimes get glossed over in organic chemistry. Nevertheless, these fundamental concepts underpin organic chemistry; without a firm understanding of these principles, organic chemistry would seem much like magic. Fundamentally, atoms and molecules combine in ways that are in accord with the laws of science. Therefore, we need to learn why atoms come together to form molecules and extend these concepts to organic chemistry. In other words, we need to be able to apply these principles to structure and function. After all, the chemical bonding between atoms makes the formation of molecules possible, from the simplest diatomic compound, H_2, to highly complex biomolecules like DNA, with molecular masses exceeding 1,000,000 grams/mole. We start at the beginning with a few basic atomic properties.

2.2. Atomic Properties

2.2.a. Atoms and the Periodic Table

Atoms are made up of three major subatomic particles - protons, neutrons, and electrons. It is the unique combination of these particles that give rise to the various elements. The properties exhibited by individual atoms (called elements on a larger scale) arise from the ratio of these subatomic particles to each other, as well as the interaction of the positively charged protons with the negatively charged electrons. Indeed, these atomic properties are responsible for how and why elements bond together to form molecules.

It is no coincidence that Mendeleev composed the periodic table as he did. Grouping atoms in periods, that is, placing atoms with like properties together, produces a table which shows trends in atomic properties. From left to right and top to bottom on the table, atomic size and number of subatomic particles increases. Starting at the bottom left of the table and moving upward and to the right, electronegativity increases, owing to the decreased atomic size and increased number of valence electrons. Within the columns or groups, atoms contain the same number of valence electrons (electrons in an atom's outermost valence shell).

Before we discuss them in more detail, however, we need to understand what it is that causes electrons, which bear the same charge, to be attracted to one another. The critical question is how the electrons interact as atoms are brought together. To clarify this, we turn to two fundamental tenets - the Pauli Exclusion Principle and Hund's Rule.

2.2.b. The Pauli Exclusion Principle and Hund's Rule – Attraction Between Electrons.

Recall from general chemistry that electrons surrounding the nucleus of an atom have energies defined generally by their distance from the nucleus. By convention, we say that electrons occupy specific "orbitals" that differ in size and shape according to their energy.

Before we go much further, let's discuss the idea of orbitals. The thing to understand is that atomic orbitals (and as we shall see later, molecular orbitals) are mathematical constructs only. Electrons are not "found" anywhere nor do they occupy any certain location around the nucleus of an atom. In fact, the Heisenberg Uncertainty Principle states that we can never know both the precise location and the exact momentum of an electron at the same time. Therefore, an orbital at best, describes the probability of an electron's location. Should you choose to take physical chemistry at some later time, this topic will be explained in much greater detail.

Nevertheless, we can identify electrons by four quantum numbers, n, l, m_l, and m_s. They define the energy states and orbitals available to an electron – an address if you will. Quantum numbers are described below.

- *n* (principal quantum number) – defines the energy level of the electron
- *l* (angular momentum quantum number) – defines the shape of the orbital where the electron "resides", often called the subshell
- m_l (magnetic quantum number) – defines the orientation of the orbital in the x, y, z plane
- m_s (spin quantum number) – defines the spin orientation of the electron

Electrons found in the outermost orbital shell (as defined by the principal quantum number, *n*) are said to be valence electrons. We concentrate on valence electrons, for they are farthest from the positively charged nucleus and thus the valence electrons are usually involved in the formation of chemical bonds.

The Pauli Exclusion Principle says that no two electrons can have the same four quantum numbers. Electrons in the same subshell may have the same *n*, *l*, and m_l quantum numbers, but they will have a different spin quantum number. By virtue of the angular momentum generated by their movement, electrons spin in a clockwise or counterclockwise direction, giving rise to m_s values. By convention, we say the spin is either up (α) or down (β).

Put another way, electrons with the same spin cannot occupy the same orbital simultaneously. This is expressed clearly by Hund's Rule which states that maximum stability is achieved by placing spin-aligned electrons in separate orbitals or domains of equal energy. Thus, a maximum of two, spin-paired electrons can occupy each orbital. Let's look at an example – the valence shell ground state orbital diagram of the carbon atom – that is, the most stable state. Remember, the boxes shown in Figure 2.1 below only represent our attempt at bookkeeping. They are our invention – the atom does not have boxes containing electrons.

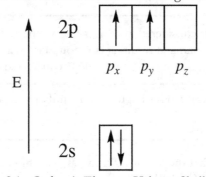

Figure 2.1. Carbon's Electron Valence Shell Ground State.

Notice that the maximum number of electrons (two) with opposite (antiparallel) spins occupy the 2s orbital, in agreement with the Pauli Exclusion Principle. The remaining two valence electrons having parallel spins are located in separate p orbitals in accordance with Hund's Rule. Bond formation between atoms uses a similar arrangement as the electrons in the 2s orbital; that is, a pair of electrons with spins aligned antiparallel. That's the picture at the atomic level. Why does this affect bonding between atoms? Let's have a closer look at Lewis symbols.

2.2.c. Lewis Symbols – Depiction of Valence Shell Electrons

Lewis symbols are graphical descriptions of atoms and their valence shell electrons. Learning to draw these is vital to the understanding of organic chemistry because Lewis symbols may be assembled into molecular Lewis structures.

To draw the Lewis symbol for an atom such as sodium (Na), we must determine the number of electrons in the outer shell (the valence electrons). These are readily discerned by locating the atom's position on the periodic table. Thus, Na is found in Group 1A and possesses one 3s electron. The symbol is depicted as Na•. In the Lewis structure, the Na part stands for the sodium nucleus

plus the electrons that are nonvalence electrons. The single dot after Na symbolizes the lone valence electron found in the 3s subshell. For atoms such as carbon (group 4A) and fluorine (group 7A), which contain larger numbers of valence electrons, we place the electrons around the atomic symbol until we have one electron on each side of the atomic symbol. Additional electrons are paired with those until we supply all valence electrons:

$$\cdot \overset{\cdot}{\underset{\cdot}{C}} \cdot \qquad\qquad\qquad \cdot \overset{\cdot\cdot}{\underset{\cdot\cdot}{F}} :$$

one electron is placed on each   electrons are paired in accord with
side in accord with Hund's Rule   the Pauli Exclusion Principle

Drawing these depictions of atoms accurately is important because we build molecules from Lewis symbols. In addition, the number of unpaired valence electrons for an atom tells us how many bonds it can form with other atoms. We'll examine this in more detail in a moment.

2.2.d. Electronegativity – Attraction Between Atoms and Electrons in Chemical Bonds

All atoms attract electrons when placed in proximity to other atoms (as they would be in a chemical bond). The extent of this attraction is called electronegativity. In general, the higher the electronegativity, the greater the attraction an atom will have for electrons in a bond with other atoms. As we shall see, the fact that different atoms have substantially different electronegativities gives rise to different types of bonding properties.

By examining differences in the electronegativities (ΔE_n) between two atoms, we can predict the type of bonding that a compound will exhibit. Three general types of chemical compounds emerge from ΔE_n calculations such that:

- if $\Delta E_n \geq 2.0$, the bond is considered ionic as is the compound;
- if $2.0 > \Delta E_n \geq 0.3$, the bond and resulting compound is polar covalent;
- if $\Delta E_n < 0.3$, the bond and resulting compound is considered covalent.

Let's use our problem solving approach from Chapter 1 to identify the type of bonding found in some example cases.

Example Problem 1. Predict whether the following compounds exhibit ionic, polar covalent or covalent bonding: (a) Na-F   (b) H_3C-Cl   (c) H_3C-CH_3

Given: (a) Na-F   (b) H_3C-Cl   (c) H_3C-CH_3
Find: Type of bonding (ionic, polar covalent or covalent) in the given molecules
Plan: Use ΔE_n to determine the degree of polarity in the bonds indicated; apply the ΔE_n magnitude rules to arrive at the type of bonding.
Solve:
(a) Na-F: ΔE_n (F) – ΔE_n (Na) = 4.0 – 0.9 = 3.1, $\therefore$ ionic
(b) H_3C-Cl: ΔE_n (Cl) – ΔE_n (C) = 3.0 – 2.5 = 0.5, $\therefore$ polar covalent
(c) H_3C-CH_3: ΔE_n (C) – ΔE_n (C) = 2.5 – 2.5 = 0.0, $\therefore$ covalent
Check: ΔE_n calculated properly, rules applied correctly.

These three categories of bonding are important and bear greatly on all fields of chemistry. We will explore these bonding types shortly.

Problem 2.1. Classify the following bonds as ionic, polar covalent, covalent. Using the method described in section 2.2b, rank order them from most (1) to (4) least ionic in nature.

(a) C-Cl bond in CH₃-Cl (b) N-H bond in NH₃ (c) Li-F bond (d) Si-CH₃ bond in Si(CH₃)₄.

2.3. Bonding Theories

Most students first learn of bonding in general chemistry when they are introduced to **v**alence **s**hell **e**lectron **p**air **r**epulsion (VSEPR) theory, **v**alence **b**ond (VB) theory and to a lesser degree, **m**olecular **o**rbital (MO) theory. Organic chemistry texts address bonding in organic molecules assuming that you have had this introduction. Depending on the text, VSEPR, VB and MO theories are used qualitatively to describe the sharing of electrons between atoms, how molecules form from atoms, and authors often invoke them to explain molecular geometry. Unfortunately, most students leave this part of an organic chemistry course confused, believing that one or more or the theories arise from the others. Although these three theories have their basis in mathematics, they are not derived from one another. Students who do not grasp this important fact will have difficulty understanding the different information each theory provides. Let's examine each in turn and highlight the information that we can get from VSEPR, VB and MO theories.

2.3.a. VSEPR Theory

Using the carbon atom as an example, according to VSEPR theory, the most probable distribution of the four valence electrons (all having the same spin, let's call it α, denoted by •) around the nucleus will assume a tetrahedral arrangement to maximize the distance and therefore, the separation between them. As Figure 2.2 shows, if we bonded the carbon atom to four hydrogen atoms (each with a β electron, denoted by ○) to make methane (CH₄), the bonds formed would also be in a tetrahedral arrangement, so as to minimize the repulsion between the electron pairs.

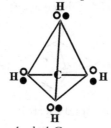

Figure 2.2. Tetrahedral Geometry of Methane.

We depict bonding pairs of electrons by either explicitly drawing a pair of electrons between atoms (Lewis electron-dot structure) or by drawing a line (Kekulé or line-bond structure). We'll cover how to draw those in the next chapter.

When lone pairs replace bonding pairs in the VSEPR model, the geometry changes slightly to account for the increased repulsion between the nonbonding electron pairs. This becomes apparent when we examine the bond angles of two simple molecules, ammonia (NH₃) and water (H₂O) and compare them to methane. See Figure 2.3.

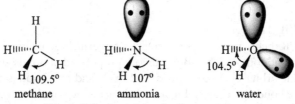

Figure 2.3. Bond Angle Comparisons.

In the latter two molecules, the nonbonding electron pair domains or orbitals take up a larger volume of space than the bonding domains defined between two nuclei. The net result is bond angle compression caused by the repulsion between the nonbonding electron pairs and bonding electron pairs.

Problem 2.2. What would be the expected C-O-H bond angle for methanol, given the line-bond structure below:

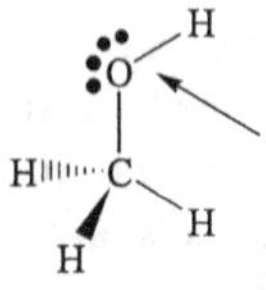

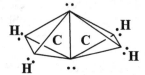

methanol

VSEPR theory also permits us to predict the geometry of multiply bonded molecules. In the case of ethene, C_2H_4, VSEPR theory proposes an electron pair arrangement as shown below:

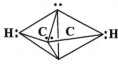

Figure 2.4. Double Bond Electron Pair Arrangement.

Thus, bond formation in ethene leads to H-C-H bond angles of approximately 120°, once again to achieve maximum separation of the bonding electron pairs. This results in a planar arrangement of the carbon-hydrogen bond network. A similar argument can be made for triple bonds, where the electron pair arrangement results in H-C-C bond angles of 180°, and thus, a linear geometry. See Figure 2.5.

Figure 2.5. Triple Bond Electron Pair Arrangement.

Cases for molecules with multiple bonds containing atoms with nonbonding electron pairs are applied in the same way as shown above in Figures 2.3-2.5.

> In summary, VSEPR gives us two important pieces of information:
> ✓ a bonding description for simple molecules, and
> ✓ a prediction of the geometry of simple molecules.

It does not supply us with all the information we need to get a complete picture of bonding in a molecule, but it is a good start.

2.3.b. Valence Bond (VB) Theory

Whereas VSEPR theory permits us to predict a molecule's geometry, VB theory only gives a very simple bonding description. Valence Bond theory proposes that bonds are formed by overlap of singly occupied atomic orbitals between atoms, and that bond formation is always in the direction of greater overlap of the bonding orbitals. Figure 2.6 depicts how s and p atomic orbitals may overlap to form sigma (σ) and pi (π) bonds, respectively. Sigma bonds form as a result of direct, head-on overlap of orbitals, while pi bonds form from sideways overlap of atomic orbitals. In the case of the p orbital, the dark and light shaded regions arise from the different algebraic signs of the atomic wave functions which describe them.

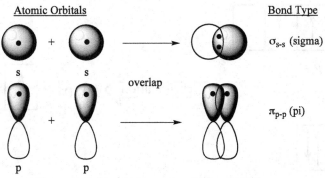

Figure 2.6. Bond Formation Using VB Theory.

While this model is useful when considering bonding between two similar orbitals, it cannot explain the geometry that is observed in cases where bonding between atoms involves different orbital sets.

Let's go back to methane. The atomic orbitals available to carbon for bonding are the 2s, and three 2p orbitals. According to VB theory, we distribute the four valence electrons of carbon so that each orbital is singly occupied. The hydrogens have only the singly occupied 1s atomic orbital. If we bond 1 hydrogen atom using its 1s orbital electron to the carbon, we have two choices – either bond the hydrogen to the carbon's lone 2s orbital electron or to one of the 2p orbital electrons. In the end, four C-H bonds of methane would be formed: 1 σ_{1s-2s} and 3 σ_{1s-2p} bonds.

Three problems arise from this. **1** - Sigma (σ) bonds formed from overlap of s atomic orbitals would be shorter than those formed by overlap of s and p atomic orbitals. Experimental evidence, however, shows that the four bonds of methane are of equal length. **2** – Interaction between s and p atomic orbitals does not result in good overlap, hence, a weak bond would be expected. **3** – VB theory gives no indication of the geometry the molecule would adopt, because it does not account for the repulsive interaction between bonding and nonbonding electron pairs.

> Ultimately, VB theory tells us about the:
> ✓ number of bonds which will form by overlap of atomic orbitals,
> ✓ and type of bonds which will form.

However, it must be used in conjunction with other theories to permit us to form a complete picture of bonding in molecules.

2.3.c. Hybridization

How do we reconcile that the C-H bonds in methane have been experimentally determined to be the same length, and that the molecular geometry is tetrahedral? Neither VSEPR nor VB theory can account for these characteristics fully. We must look to a different model – one that does explain the observed experimental results. In the 1930s, Linus Pauling developed the concept of hybridization, a mathematical description of how atomic orbitals can be recombined. It turns out that if one combines the atomic orbitals containing carbon's valence shell electrons to produce four equivalent "hybrid" orbitals, molecular geometry and equivalency of the bond lengths in methane are accounted for very well. Let's see how it works.

Recall from Figure 2.1 carbon's electron valence shell ground state. Now, if we "mix" the mathematical expression for the 2s and 2p electrons' atomic orbitals correctly (these are called wave functions and we'll talk about them in the next section), we arrive at 4 equivalent hybrid orbitals. Put another way, we get a slight energy savings by having all the valence electrons equivalent in energy, as Figure 2.7 shows.

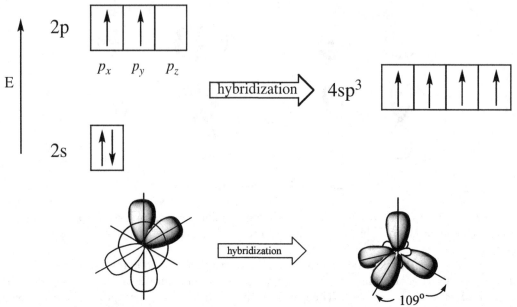

Figure 2.7. sp³ Hybridization of Carbon's Atomic Orbitals.

Combining these wave functions in this way has several advantages. First, the hybridized wave functions are directional in nature as Figure 2.7 shows. Second, the large lobe, which faces the bonding direction, has the size required to overlap very effectively with other atoms, leading to stronger bonds. Third, to minimize the interaction between electrons, the sp³ orbitals adopt a tetrahedral arrangement for bonding. Finally, because the electrons occupying the sp³ orbitals are the same energy and intermediate in energy relative to the atomic orbitals, the single (σ) bonds formed with hydrogen (if we consider methane) will have the same bond lengths.

Hybridization is easily extended to molecules with multiple bonds. In the case of double bonds, as Figure 2.8 shows, mixing of the wave functions of the 2s and two 2p electrons results in three sp² hybrid orbitals. The remaining p orbital with its unpaired electron is unhybridized and can be used to form a π bond, should the circumstances permit.

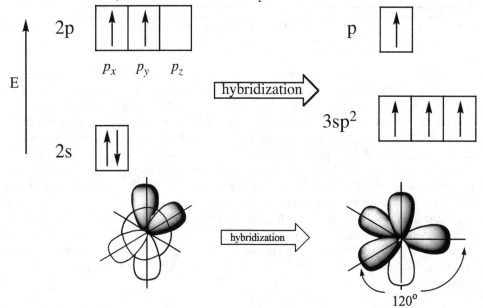

Figure 2.8. sp² Hybridization of Carbon's Atomic Orbitals.

Just as in the case of sp³ hybridization, sp² hybrid orbitals adopt a geometry that minimizes the interaction between the electrons. This results in a planar arrangement with a 120° separation between the orbitals. As our figure depicts, the p orbital lies 90° out of the plane of the sp² hybrids.

For triple bonds, Figure 2.9 indicates how mixing wave functions of the 2s electron and one 2p electron results in two sp hybrid orbitals. The two remaining p orbitals with their unpaired electrons are unhybridized and will form two π bonds.

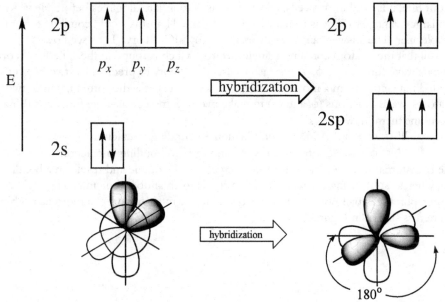

Figure 2.9. sp Hybridization of Carbon's Atomic Orbitals.

As before, sp hybrid orbitals adopt a geometry that minimizes the interaction between the electrons. This results in a linear arrangement with a 180° separation between the orbitals. As Figure 2.9 shows, one p orbital is in the same plane and the other is 90° out of the plane of the sp hybrids.

Problem 2.3. Predict the hybridization for each atom in the following structures:

methanol   acetonitrile   acetone

2.3.d. Molecular Orbital Theory

Up to now, the bonding theories we've examined are qualitative pictures of how the valence electrons of atoms may interact to form bonds. Molecular orbital (MO) theory provides us with a more complete picture of how atoms combine. Essentially, electrons in atoms may be defined by mathematical terms called wave functions. Furthermore, these wave functions are loosely depicted by the "orbitals" we have discussed thus far. They are really probability density distributions surrounding a nucleus that give us a volume of space where an electron will likely be located about 90% of the time. In other words, they provide us with a visual representation of the wave functions which actually describe the electrons.

We can't get into how these wave functions arise, but for now what you need to understand is that if we mathematically combine the atomic wave functions (ϕ) of the two atoms we wish to consider in bond formation, we arrive at an expression which describes two new molecular orbitals

(Ψ). This is a truth that will follow in every aspect of chemistry – for each pair of atomic orbitals we combine in bond formation, we generate the same number of molecular orbitals. We'll now look at two examples to help us better understand how MO theory works.

> The H_2 Story

Most all who have had any introduction to chemistry know that H_2 is a gas found in nature. It powers our Sun, it is used in fuel cells, it even was used to inflate dirigibles before He was readily available in large quantities. Why is it that two perfectly fine H atoms would combine to form a diatomic molecule? The answer can be easily explained by MO theory. Let's see how.

Recall that the H atom contains a single electron which is found in the spherical 1s orbital. We'll define its wave function as ϕ. If we wish to use MO theory to predict how two H atoms (H_a and H_b) would interact to form a molecule of H_2 gas, we can express the combination of two hydrogen atoms' wave functions (ϕ_a and ϕ_b) to make two new molecular wave functions (Ψ_1 and Ψ_2). We may combine them in two ways:

$$\Psi_1 = \phi_a + \phi_b \quad \text{(Additive combination} \rightarrow \text{bonding interaction)}$$
$$\Psi_2 = \phi_a - \phi_b \quad \text{(Subtractive combination} \rightarrow \text{antibonding interaction)}$$

Adding the two atomic wave functions is more energetically favorable and results in a bonding interaction, whereas subtracting the wave functions leads to an antibonding interaction.

To graphically depict how these atoms combine, we must construct a molecular orbital energy diagram as shown in Figure 2.10.

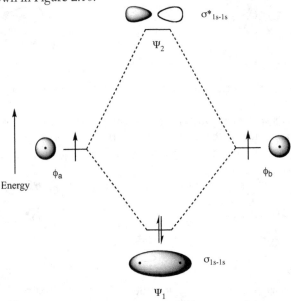

Figure 2.10. H_2 Molecular Orbital Energy Diagram.

There are several things to note about our diagram. It assumes that initially, the H atoms are at their normal energetic state and that they are too far apart to interact. As the atoms approach each other to the exact distance required for bonding, interaction or overlap of the atomic orbitals leads to creation of two new molecular orbitals, one much lower in energy than the atomic orbitals, and one very much higher in energy. In accordance with the Aufbau principle, the electrons from the hydrogen atoms occupy the lowest energy molecular orbital. Also, as the Pauli Exclusion Principle requires, the two electrons occupy this new molecular orbital with opposite spins. This results in a $\sigma_{1s\text{-}1s}$ bond being formed, where maximum overlap is achieved by the additive nature of the wave functions. Meanwhile, the antibonding molecular orbital, $\sigma^*_{1s\text{-}1s}$, which has a node signifying minimal overlap between the atomic orbitals, is left unfilled. The energy saved by the bonding of these two H atoms is an example of how a two-electron, two-MO system is stabilizing.

Problem 2.4. Prepare a molecular orbital energy diagram for the cation of diatomic hydrogen, H_2^+.

➢ The He_2 Story

Of course, not all atomic interactions are stabilizing. For example, do two He atoms combine to form He_2? The answer is no and MO theory can explain this easily if we construct a molecular orbital energy diagram.

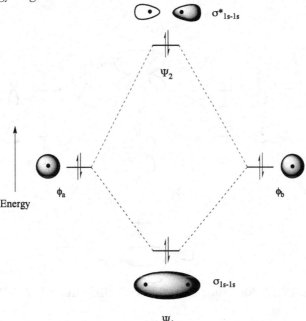

Figure 2.11. He_2 Molecular Orbital Energy Diagram.

Now it becomes clear why He_2 cannot exist. Even though two of the electrons from the He atoms produce a bonding interaction (stabilizing), the two remaining electrons are required by the Aufbau principle to occupy the antibonding orbital. Since the energy cost of placing these electrons into the antibonding orbital is greater than that saved by the bonding interaction, the net result would be a molecule that is less stable than the two He atoms. Therefore, He_2 does not form. For this reason, we say that a four-electron, two-MO interaction is destabilizing. The concepts of a stabilizing two-electron two-MO interaction and destabilizing four-electron two-MO interaction are critical to our thinking about reactivity in organic compounds and will be elaborated on later in the text.

➢ Bonding between Carbon Atoms

The single bond found in H_2 is analogous to that found in carbon-carbon σ bonds. In our MO energy diagram, we merely replace the 1s atomic orbitals with sp^3 hybrid orbitals that carbon uses in single bonds. The net result is a two-electron, two orbital interaction – a stabilizing interaction that results in formation of a $\sigma_{sp3\text{-}sp3}$ bond.

Problem 2.5. Prepare a molecular orbital energy diagram for the cation of diatomic helium, He_2^+.

Problem 2.6. Prepare a molecular orbital energy diagram for carbon-carbon bond in ethane, $H_3C\text{-}CH_3$.

MO theory also explains how carbon atoms can form multiple bonds. In Figure 2.12, we consider the formation of the carbon-carbon double bond of ethene, $H_2C=CH_2$. If we bring two sp^2 hybridized atomic orbitals together to form the carbon-carbon σ bond by head-on overlap and two unhybridized p orbitals together to form the π bond by sideways overlap, the energy diagram shows the formation and relative energies of all the molecular orbitals that result from these interactions. Note that although this is a four-electron process, there are four molecular orbitals. We also recognize that placing the four electrons found in the atomic orbitals into the two lowest energy molecular orbitals results in a bonding (stabilizing) interaction. That is, the net energy of the molecular orbitals is lower than that of the atomic orbitals we began with.

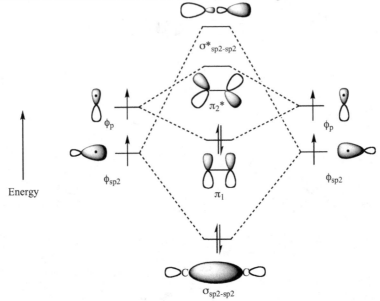

Figure 2.12. Molecular Orbital Energy Diagram for $H_2C=CH_2$.

Obviously, we may extend this to show how a triple bond forms in an analogous way. That is left as an exercise for you.

Problem 2.7. Prepare a molecular orbital energy diagram for the carbon-carbon bonds in ethyne, HC≡CH.

2.4 Types of Bonds and Compounds

As we mentioned earlier, compounds and the bonds that make them up can be roughly grouped into three categories, ionic, polar covalent, and covalent. Because ionic bonds and compounds are different in the way they occur, we'll address them first and finish up with polar covalent and covalent compounds.

2.4.a. Ionic Bonds

Now that we have looked at bonding theories, let's discuss how we classify the bonds. The guiding principle in constructing bonds comes from the tendency of atoms to acquire or approach the electron configuration associated with the noble gas; i.e. a filled outer shell. Since a filled outer shell for Ne-Rn contains eight electrons, the octet rule has emerged. In the case of He, which has two electrons in its filled outer shell, the duet rule has been formulated. Application of these rules will be illustrated in the following examples. Consider the metal-nonmetal interaction of a sodium atom and a fluorine atom using Lewis symbols:

$$Na\cdot \quad + \quad \cdot \ddot{\underset{\cdot\cdot}{F}}:$$

We would anticipate that the sodium atom would tend to lose or donate an electron readily and become Na^+ in order to achieve the electron configuration of Ne.

16

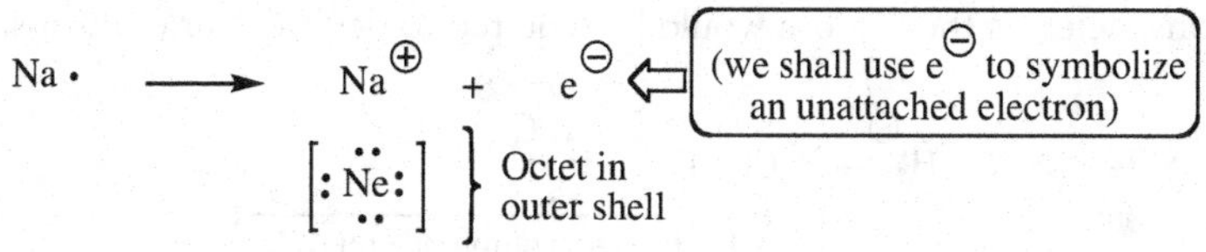

$$Na\cdot \longrightarrow Na^{\oplus} + e^{\ominus} \Longleftarrow \boxed{\text{(we shall use } e^{\ominus} \text{ to symbolize an unattached electron)}}$$

$$\left[:\overset{\displaystyle ..}{\underset{\displaystyle ..}{Ne}}:\right] \left.\right\} \begin{array}{l}\text{Octet in}\\\text{outer shell}\end{array}$$

The fluorine atom, on the other hand, would be seeking an additional electron and become F^{-} in order to acquire the electron configuration of Ne.

$$:\overset{\displaystyle ..}{F}\cdot + e^{\ominus} \longrightarrow :\overset{\displaystyle ..}{\underset{\displaystyle ..}{F}}:^{\ominus}$$

$$\left[:\overset{\displaystyle ..}{\underset{\displaystyle ..}{Ne}}:\right] \left.\right\} \begin{array}{l}\text{Octet in}\\\text{outer shell}\end{array}$$

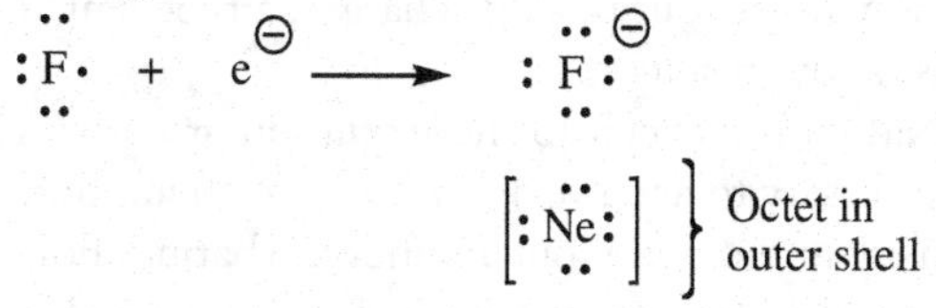

Adding these equations leads to the overall result shown below:

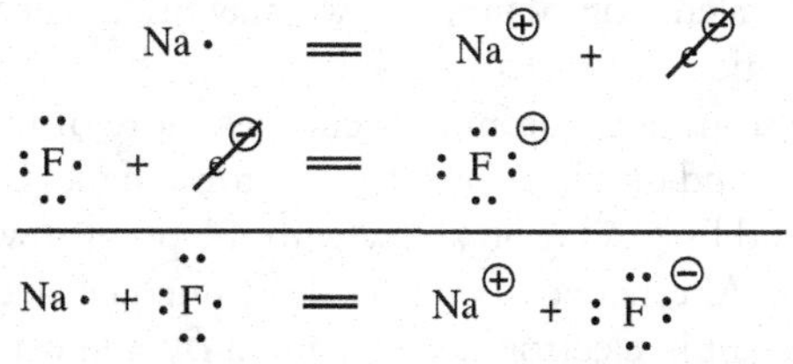

$$Na\cdot = Na^{\oplus} + \cancel{e}$$

$$:\overset{\displaystyle ..}{F}\cdot + \cancel{e} = :\overset{\displaystyle ..}{\underset{\displaystyle ..}{F}}:^{\ominus}$$

$$\rule{6cm}{0.4pt}$$

$$Na\cdot + :\overset{\displaystyle ..}{\underset{\displaystyle ..}{F}}\cdot = Na^{\oplus} + :\overset{\displaystyle ..}{\underset{\displaystyle ..}{F}}:^{\ominus}$$

By the transfer of an electron from the sodium atom to the fluorine atom both the resulting ions have realized the filled shell electron configuration of Ne since each ion has eight electrons in the outer shell. The positive charge on the product sodium ion appears since a negatively charged electron has been lost from the neutral atom. The product fluoride ion displays a negative charge since the neutral fluorine atom has acquired a negatively charged electron. The opposite charges attract and a bond is formed. Since the bonding involves oppositely charged ions, the result is aptly termed an ionic bond.

Examining the electronegativities of the atoms is another way of arriving at the same conclusion. If the difference in electronegativity between the atoms (ΔE_n) is large enough ($\Delta E_n > 2$), we classify the bond formed as ionic. The atom with the larger electronegativity (F) has a greater affinity for electrons and will accept the other atom's electron in the process of bond formation. In effect, it becomes an anion. The less electronegative atom (Na) has a lower ionization potential and can be induced to donate its electron to form part of the ionic bond, essentially becoming a cation. In essence, bonding results from the electrostatic attraction between the differently charged ions.

This type of bonding is critical for the formation of water-soluble compounds, such as various salts (NaCl), acids (HCl), and bases (NaOH). Their ability to dissolve in a variety of polar solvents permits the ions to participate in numerous reactions. Without ionic bonding, a host of chemical processes in the body would stop since the body is nearly 70% water. Ionic compounds also are reactants in many organic reactions; their solubility in aqueous or alcoholic solutions and nonprotic, polar solvents enables these reactions to take place.

2.4.b. Polar Covalent Bonds

Let us now examine a different situation. Consider the hypothetical process below, depicting the formation of trichloromethane ($CHCl_3$).

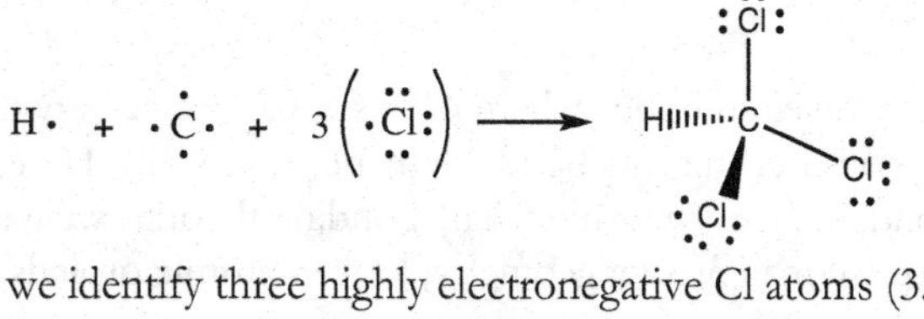

$$H\cdot + \cdot \overset{\displaystyle .}{\underset{\displaystyle .}{C}}\cdot + 3\left(\cdot\overset{\displaystyle ..}{\underset{\displaystyle ..}{Cl}}:\right) \longrightarrow \begin{array}{c} :\overset{\displaystyle ..}{Cl}: \\ | \\ H\!\!\text{\tiny IIIIII}C \\ \end{array}$$

In this equation, we identify three highly electronegative Cl atoms (3.0), and the more moderately electronegative C (2.4) and H (2.1). Based on our rationale in the previous section, each Cl atom and the H atom would be expected to accept an additional electron to complete their noble gas configurations, whereas the C atom would be required to donate four electrons. Here is the result:

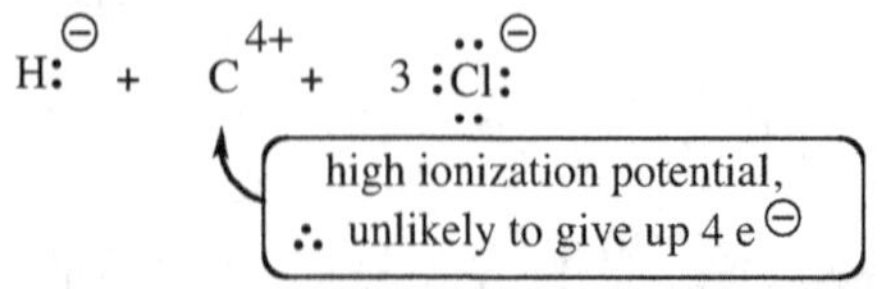

$$H\overset{\ominus}{:} \quad + \quad C^{4+} \quad + \quad 3 \; :\overset{..}{\underset{..}{Cl}}\overset{\ominus}{:}$$

$$\boxed{\begin{array}{c} \text{high ionization potential,} \\ \therefore \text{ unlikely to give up 4 e}^{\ominus} \end{array}}$$

Therefore, while both the H and Cl atoms can achieve noble gas configurations by accepting an additional electron, carbon now bears four positive charges, a repellent and unlikely result. Why does this method fail to yield satisfactory results?

Unlike the case of NaF, it is difficult for nonmetals in groups 3A-5A to readily donate or receive the 3 – 5 electrons necessary to achieve a noble gas configuration. In the case of HCCl$_3$, we have sharing of electron density in the bond, but an uneven sharing. For atoms involved in bonds where ΔE_n is < 2 but ≥ 0.3, we classify the bond formed as polar covalent. The bonding theories we have discussed apply in polar covalent compounds, so we may predict geometry, look at atomic orbitals and talk about the formation of molecular orbitals.

In effect, the more electronegative atom concentrates more of the electron density in the bond near itself. This creates a bond dipole. Consider chloromethane, CH$_3$Cl. We show a net dipole moment for the C-Cl bond by placing an arrow with the point toward the more electronegative atom – chlorine. Alternatively, we may show unequal electron distribution in this bond by indicating centers deficient in electron density with a δ^+, and centers with excess electron density with a δ^-. Although there is a modest ΔE_n between the C and H atoms, these are not substantial and can usually be discounted when determining the net dipole moment for the molecule.

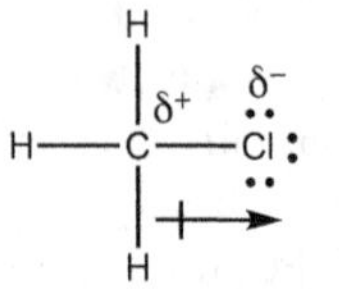

$$\begin{array}{c} H \\ | \quad \overset{\delta^-}{\underset{..}{\overset{..}{Cl}}} \\ H\!-\!\overset{\delta^+}{C}\!-\!Cl: \\ | \\ H \end{array}$$

Figure 2.13. Chloromethane's Dipole Moment.

However, the presence of polar covalent bonds in a molecule does not automatically make a molecule polar. As an example, consider carbon tetrachloride, CCl$_4$, with its four C-Cl bonds:

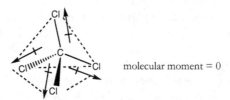

molecular moment = 0

Figure 2.14. Carbon Tetrachloride's Dipole Moment.

In this case, the individual bond moments, because of tetrahedral geometric symmetry, sum to give a net molecular dipole moment of 0. Therefore, CCl$_4$ is nonpolar.

Problem 2.8. Predict the net dipole moment of dichloromethane, CH$_2$Cl$_2$.

2.4.c. Covalent Bonds

For those atoms involved in bonds where ΔE_n is < 0.3, we classify the bond formed as covalent. As an example, consider the C-C bond in ethane, H$_3$C-CH$_3$. Here, we have sharing of electron density in the bond, in an even fashion. The bonding theories we have discussed also apply in covalent compounds, so we may predict geometry, look at atomic orbitals and talk about the formation of molecular orbitals.

2.5 Summary

Atomic properties dictate how atoms interact with each other. From knowledge of electronegativity, application of the Pauli Exclusion Principle, Hund's Rule, and the theories of bonding, we are now able to predict not only how atoms will bond to form many organic molecules, we can also readily discuss the geometry of these molecules. We are ready for our next big step, taking this knowledge and using it to draw organic molecules with structural and geometric accuracy.

Solved Problems

2.1.
Given: (a) C-Cl bond in CH_3-Cl (b) N-H bond in NH_3 (c) Li-F bond (d) Si-CH_3 bond in $Si(CH_3)_4$.
Find: Classify the following bonds as ionic, polar covalent, covalent. Rank order them from most (1) to (4) least ionic in nature.
Plan: Use ΔE_n to determine the degree of polarity in the bonds.
Solve: (a) ΔE_n (C-Cl) = 3.0-2.5=0.5, $\therefore$ polar covalent.
(b) ΔE_n (N-H) = 3.0-2.1=0.9, $\therefore$ polar covalent.
(c) ΔE_n (Li-F) = 4.0-1.0=3.0, $\therefore$ ionic.
(d) ΔE_n (Si-C) = 2.5-1.8=0.7, $\therefore$ polar covalent.
Check: (1) Electronegativities determined correctly, (2) ΔE_n calculations performed correctly.

2.2.
Given: Structure for methanol:

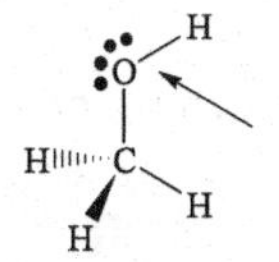

methanol

Find: The expected C-O-H bond angle for methanol.
Plan: Apply VSEPR theory for bond angle estimation.
Solve: sp^3 hybridized atoms (109.5°) with 2 bonds and 2 electron pairs (-2.5° each):
$$109.5° - 2(2.5°) = 104°.$$
Check: (1) Hybridization determined correctly from VSEPR, (2) Bond $\angle$ calculations performed correctly.

2.3.
Given: Structures below:

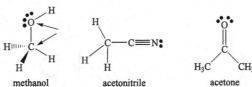

methanol   acetonitrile   acetone

Find: Predict the hybridization for each atom.
Plan: Apply VSEPR theory: (a) Count the bonds and non-bonding electron pairs around each atom and (b) consider the bonding arrangements of each atom.
Solve: Methanol: C: 4 single bonds, $\therefore$ sp^3; O: 2 bonds and 2 electron pairs, $\therefore$ sp^3.
Acetonitrile: H_3**C**-C: 4 single bonds, $\therefore$ sp^3; C-**C**≡N: 1 single bond and 1 triple bond, $\therefore$ sp; N: 1 non-bonding electron pair and 1 triple bond, $\therefore$ sp.
Acetone: H_3**C**-C: 4 single bonds, $\therefore$ sp^3; C-**C**=O: 2 single bonds and 1 double bond, $\therefore$ sp^2; O: 2 non-bonding electron pairs and 1 double bond, $\therefore$ sp^2.
Check: Hybridization determined correctly from VSEPR.

2.4. Prepare a molecular orbital energy diagram for the cation of diatomic hydrogen, H_2^+.
Given: Cation of diatomic hydrogen, H_2^+.
Find: Molecular orbital energy diagram.
Plan: (a) Using MO theory, construct a relative energy diagram depicting the atomic orbitals (AOs) that interact to produce the MOs.
Solve:

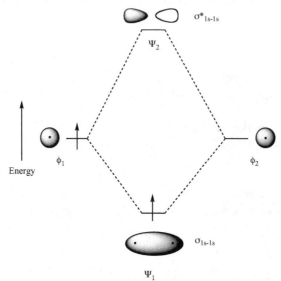

Check: (1) AOs drawn correctly, (2) Interaction b/w AOs determined, (3) MOs formed are included in the diagram, (4) Electrons are placed in MOs correctly.

2.5. Prepare a molecular orbital energy diagram for the cation of diatomic helium, He_2^+.
Given: Cation of diatomic hydrogen, He_2^+.
Find: Molecular orbital energy diagram.
Plan: (a) Using MO theory, construct a relative energy diagram depicting the atomic orbitals (AOs) that interact to produce the MOs.
Solve:

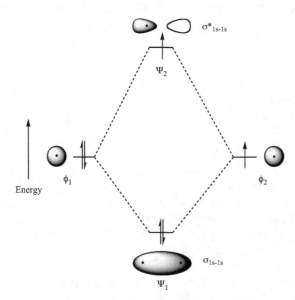

Check: (1) AOs drawn correctly, (2) Interaction b/w AOs determined, (3) MOs formed are included in the diagram, (4) Electrons are placed in MOs correctly.

2.6. Prepare a molecular orbital energy diagram for the carbon-carbon bond in ethane, H_3C-CH_3.
Given: Ethane, H_3C-CH_3.
Find: Molecular orbital energy diagram for the carbon-carbon bond.
Plan: (a) Using MO theory, construct a relative energy diagram depicting the hybrid orbitals that interact to produce the MOs.
Solve:

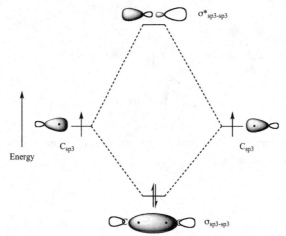

Check: (1) Hybrid orbitals drawn correctly, (2) Interaction b/w hybrid orbitals determined, (3) MOs formed are included in the diagram, (4) Electrons are placed in MOs correctly.

2.7. Prepare a molecular orbital energy diagram for the carbon-carbon bonds in ethyne, HC≡CH.
Given: Ethyne, HC≡CH.
Find: Molecular orbital energy diagram.
Plan: (a) Using MO theory, construct a relative energy diagram depicting the hybrid orbitals that interact to produce the MOs.
Solve:

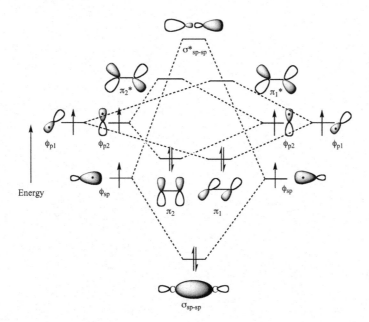

Check: (1) Hybrid orbitals drawn correctly, (2) Interaction b/w hybrid orbitals determined, (3) MOs formed are included in the diagram, (4) π MOs are degenerate (energies of π_1 and π_2 are equal), π^*MOs are degenerate (energies of π^*_1 and π^*_2 are equal), (5) electrons are placed in MOs correctly.

2.8.
Given: CH_2Cl_2.
Find: Predict the net dipole moment of dichloromethane.
Plan: (a) Draw the 3-D structure of CH_2Cl_2, (b) Determine bond dipoles, (c) Estimate the net molecular dipole moment.
Solve:

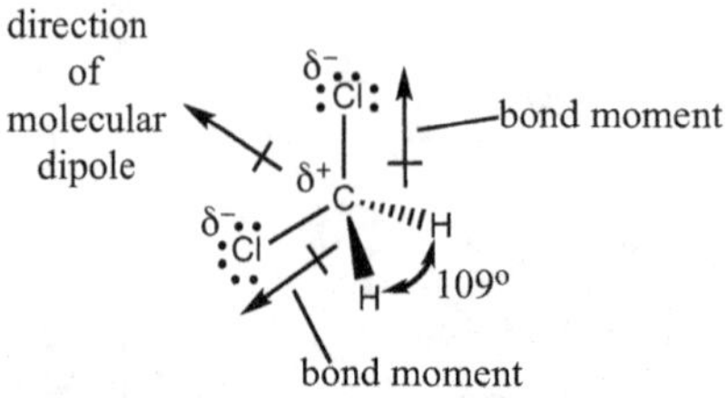

Check: (1) Structure drawn correctly, (2) Bond ∠s determined, (3) Bond moments determined, (4) Net molecular dipole moment estimated.

22

Chapter 3 – Drawing Chemical Structures

One of the most common problems for beginning organic chemistry students is drawing chemical structures that adhere to geometry considerations based on the theories we covered in Chapter 2. In addition to geometric requirements, calculation of the formal charge that an ion or molecule may possess is also necessary to gain a complete picture of the structure. Unless resolved early, the inability to translate chemical formulae into three-dimensional representations of molecules as Lewis, Kekulé or skeletal structures has serious consequences. If students cannot accurately draw organic chemical structures, knowledge of how and why molecules react the way they does will remain unclear. In other words, understanding the function of a species is not possible without knowledge of structure.

Chemical structure drawing does not have to be difficult, and if you remember the rules we discussed in chapter 2 and what we are about to cover in this chapter, you can accurately draw 2-dimensional or 3-dimensional structures which represent molecules and you can correctly account for the molecular geometry. Since bonds are formed by valence shell electrons, any structure we wish to draw must account for those electrons. To ensure this, we must do some electronic bookkeeping.

3.1. Formal Charges and Lewis Electron Dot Structures – Electronic Bookkeeping

When neutral atoms bond together, the sharing of electrons can lead to some of the atoms having more or fewer electrons in its valence shell than when it is an isolated atom. We determine this by calculating the formal charge – the charge on a particular atom within a molecule or ion. Formal charge on an atom is defined as = the # of electrons a neutral atom has in its valence shell – ½ the # of bonding electrons the atom has in a molecule or ion – the # of nonbonding electrons. Correctly drawn chemical structures must include these charges.

Let's look at a simple example – methane, CH_4. As the chemical formula shows, methane contains one atom of carbon and four atoms of hydrogen. The molecular formula provides us with one other important piece of information – how the atoms are attached. We read this formula as four hydrogen atoms - each bonded to the carbon atom. If we look at the position of carbon in the periodic table, it makes sense that carbon would be the central atom in methane for two reasons. First, carbon contains four valence electrons which are available for sharing in some types of bonds with other atoms. Secondly, carbon is not very electronegative and therefore the valence electrons would be available for sharing. We may arrange the atoms by combining the Lewis symbols for the atoms thusly in a Lewis electron dot structure:

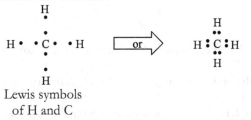

Lewis symbols
of H and C

Figure 3.1. Lewis Electron Dot Structure for Methane.

Calculation of the formal charge on C: $4 – \frac{1}{2}(8) – 0 = 0$, while H: $1 – \frac{1}{2}(2) – 0 = 0$.

Formation of the carbon – hydrogen bond has benefits. Notice that carbon in the methane molecule now has its octet completed and each hydrogen has its duet completed. For both atoms, bonding fills their valence shells. As we learned in chapter 2, this is typically what we call a covalent bond.

3.2. Kekulé or Line Bond Structures of sp³ Hybridized Molecules

Although Lewis electron dot structures clearly show how the valence electrons are utilized, depicting each of these electrons quickly becomes cumbersome in all but the simplest molecules. We need a better system, and the Kekulé structure combines simplicity with completeness, by using a single line in place of an electron pair or bond. Using this line-bond method, methane can be drawn this way:

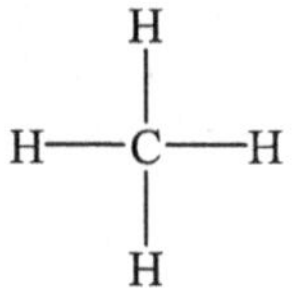

$$H-\overset{\displaystyle H}{\underset{\displaystyle H}{C}}-H$$

Figure 3.2. Kekulé or Line Bond Structure for Methane.

While we may be satisfied with our artistic prowess, this structure is somewhat misleading in that the bond angles appear to be 90°. Moreover, the drawing shown is only two-dimensional. We know that molecules are 3-dimensional. For methane, the electron pairs and the molecule should have tetrahedral geometry due to sp³ hybridization, which leads us to conclude that the bond angles should be close to 109.5°. To show this, we need to think $\underline{\text{and}}$ draw 3-dimensionally. We use drawings that contain dashed wedges and solid wedges or dashes and wedges to indicate the 3-dimensional nature of a molecule:

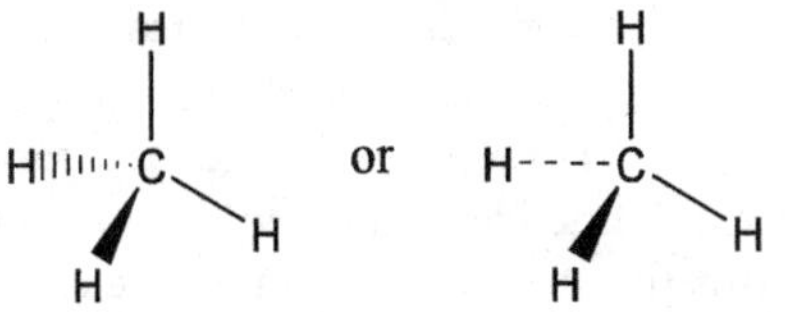

Figure 3.3. 3-Dimensional Kekulé or Line Bond Structure of Methane.

In these drawings, the wedged-H can be thought of as above the plane of the paper, while the dashed wedge-H or dashed-H is projected below the plane of the paper. The remaining two hydrogens and the carbon are in the plane of the paper.

What about molecules with atoms other than C and H? As an example, let's draw chloromethane, CH_3Cl. The only difference is that we must account for the nonbonding electron pairs (3) which chlorine contains. The structure is shown below:

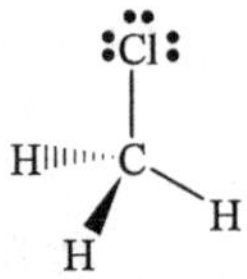

Figure 3.4. 3-Dimensional Kekulé Structure for Chloromethane.

Charged species are drawn in the same way. The key is determining the correct formal charge and indicating it in the Lewis electron dot or line-bond structure. Consider methoxide, CH_3O^- and methylammonium, $CH_3NH_3^+$. Calculation of formal charges:

	methoxide	methylammonium
C (all):	$4 - \frac{1}{2}(8) - 0 = 0$	$4 - \frac{1}{2}(8) - 0 = 0$
H (all):	$1 - \frac{1}{2}(2) - 0 = 0$	$1 - \frac{1}{2}(2) - 0 = 0$
O:	$6 - \frac{1}{2}(2) - 6 = -1$	
N:		$5 - \frac{1}{2}(8) - 0 = +1$

The 3-dimensional line-bond (Kekulé) structures are presented below. The formal charges are placed in close proximity to the atoms which bear them. We can enclose the charges with a circle to clearly identify them.

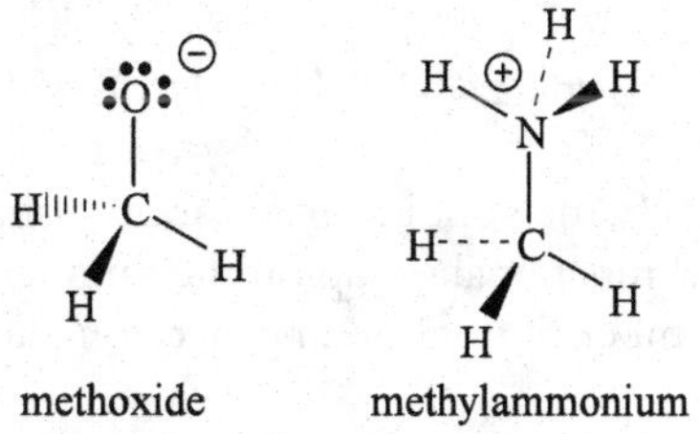

methoxide   methylammonium

Figure 3.5. Kekulé Structures for Methoxide and Methylammonium.

Problem 3.1. Using your model kit, construct models of the ions in Figure 3.5.

For larger molecules, drawing all the carbons and hydrogens bonded in a structure can get tedious. Consider 2-methylhexane, $CH_3(CH_2)_3CH(CH_3)CH_3$. The Kekulé structure we draw is:

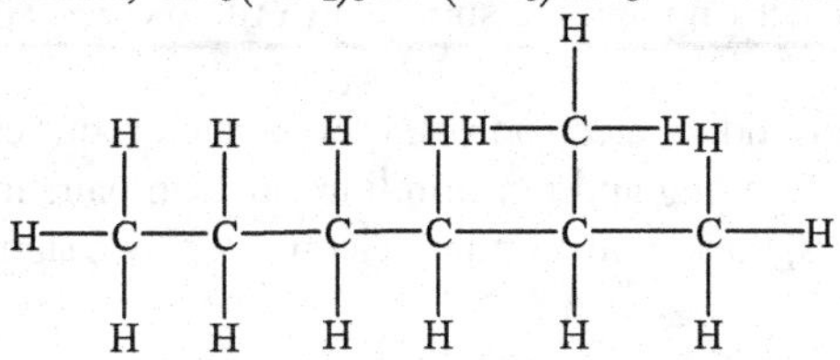

Figure 3.6. Kekulé Structure for 2-Methylhexane.

What we really mean is this:

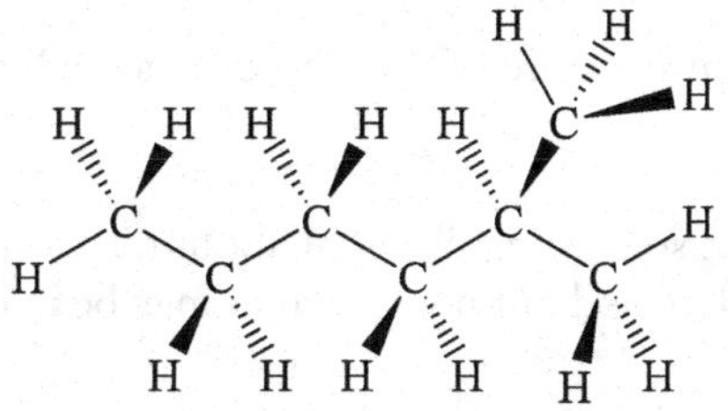

Figure 3.7. 3-Dimensional Line Bond Structure of 2-Methylhexane.

Notice the general line traced by the carbon atoms; it has a zig-zag appearance. We have drawn what is referred to as a sawhorse projection. The bond angles are reasonably close to 109.5°, and the overall tetrahedral geometry is apparent. Over the years, organic chemists have simplified drawing chemical structures to only include the carbon skeletal backbone:

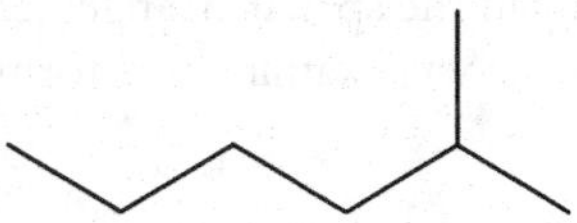

Figure 3.8. Carbon Skeletal Structure of 2-Methylhexane.

The good thing about this system is that (1) the correct molecular geometry is taken into consideration, (2) functional groups can always be depicted, and (3) hydrogens are understood but not drawn in so as to keep the time required and clutter to a minimum.

Practice Problem 1. Draw the skeletal structure for the molecule, 2-pentanol, $CH_3(CH_2)_2CH(OH)CH_3$.

Given: Expanded molecular formula: $CH_3(CH_2)_2CH(OH)CH_3$
Find: Skeletal Structure
Plan: (1) Draw the 5 carbon framework in zig-zag form, using the correct bond angles.
(2) Place the hydroxyl functional group on the number 2 carbon.
(3) Supply correct number of nonbonding electron pairs around oxygen.
Solve:

Check: Carbon framework is oriented correctly with accurate bond angles, hydroxyl functional group is placed on the correct carbon with accurate bond angle, the correct number of nonbonding electron pairs are supplied to the oxygen atom.

Here, we have depicted the nonbonding electron pairs the oxygen requires to complete its octet; it is good practice to show both the bonding and nonbonding electron pairs in chemical structures. As we will see in upcoming chapters, these non-bonding electrons play a significant role in chemical reactions.

Problem 3.2. Calculate the formal charge on the oxygen of (a) 2-pentanol and (b) the anion of 2-pentanol, 2-pentoxide.

Problem 3.3. Expand the skeletal structure of 2-pentanol to a sawhorse projection depicting all hydrogens.

Now you say that there is something different about the carbon bearing the hydroxyl group. The hydrogen that is understood but not drawn in may be in one of two positions relative to the –OH group:

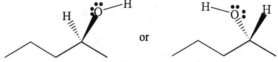

Figure 3.9. Isomeric Structures of 2-Pentanol.

These isomers of 2-pentanol differ only in the spatial arrangement of the –H and –OH functional groups; they are called stereoisomers. We will examine this form of isomerism in chapter 5.

3.3 Newman Projections

Another way of drawing the structures we have been investigating is by depicting them in a head-on sense – the Newman projection. This rendering allows us to look at a particular molecule while viewing it from the vantage point of the carbon-carbon bond. Let's look at the simplest alkane for which different conformations exist as an example: ethane.

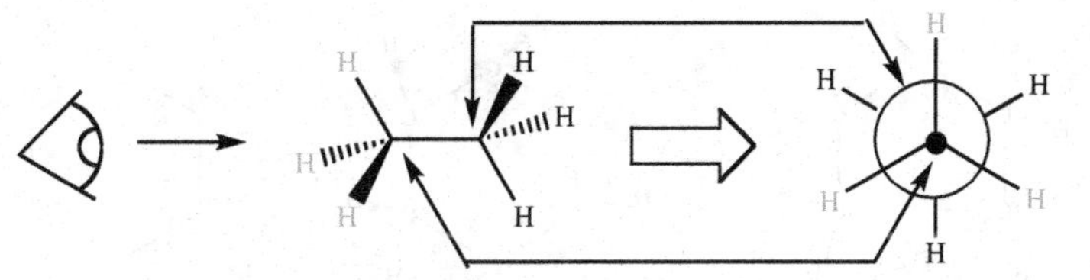

"sawhorse" projection   Newman projection

Figure 3.10. Newman Projection of Ethane.

The left structure in Figure 3.10 is a typical sawhorse projection of ethane. When we sight down the carbon-carbon bond, what we see is the 1-carbon in front as a dot in the middle of a circle which represents the 2-carbon. Looking end-on, as shown in Figure 3.10, we show the hydrogens attached to the carbons at 120° angles from each other. When we add the hydrogens on the 2-carbon, we can place them in between the hydrogens we have attached to the 1-carbon, thus maximizing the distance between them. This is called the **staggered** conformation, one that allows for the minimum interaction of atoms on adjacent carbon atoms.

Molecular orbital theory also provides an explanation as to the preference for the staggered conformation of alkanes. If we use the most stable conformation of ethane, the "sawhorse" shown in Figure 3.10, we may draw another structure (shown in Figure 3.11 on the right) depicting the antibonding molecular orbital for the C_1-H bond ($\sigma*_{C-H}$) and the bonding molecular orbital for the C_2-H bond (σ_{C-H}).

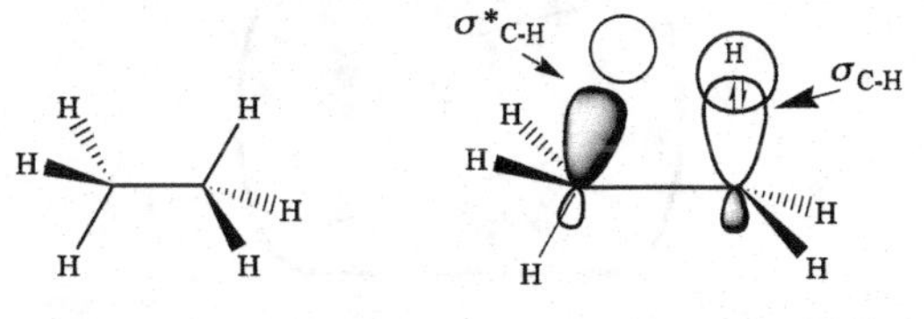

"sawhorse" projection $\sigma_{C-H} - \sigma*_{C-H}$ interaction

Figure 3.11. Stabilization of Staggered Conformation of Ethane by MO Interaction.

For ethane in this conformation, these molecular orbitals align such that a **two-orbital, two-electron interaction** takes place. Recall from Chapter 2 that this type of interaction is favorable and stabilizing. This form of stabilization, called hyperconjugation, will be discussed in Chapter 4.

Problem 3.4. Make a model of ethane and orient it to give a view similar to that of a Newman projection.

The Newman projection is a very powerful tool for examining virtually any organic molecule. As we shall see in Chapter 4, there are other conformations that are less stable than the staggered conformation. The ability to draw Newman projections will allow us to study the stability of various conformations of organic molecules in detail. We'll discuss these topics in the next chapter.

3.4 Skeletal Structures of Cyclohexanes

There are occasions when C-C bonding can lead to a cyclic array and when the necessary hydrogens are added to satisfy the octet rule, we have a member of the cycloalkane family of molecules. Although there are cycloalkanes with as few as three carbons, here we focus on the six-membered cycloalkane, cyclohexane. We will revisit other cycloalkanes in chapter 4.

Found in important naturally occurring molecules like cholesterol, cyclohexane is arguably the most common cycloalkane. Thus, we need to be able to draw it correctly. Shown below are the Kekulé, dashed-wedged and skeletal structures of cyclohexane:

Figure 3.12. Kekulé, Dashed-Wedged and Skeletal Structures of Cyclohexane.

Just as the Kekulé structure of methane fails to depict the three-dimensional nature of the molecule, these drawings have limitations. Even when we show the C-H bonds with wedges and dashes, we still do not get a true appreciation for the geometry the molecule really adopts. Over the years, the skeletal structure has been used to depict the C-C bond arrangement in cycloalkanes in the same way as in acyclic alkanes.

What we need is a fresh perspective. If we could view cyclohexane slightly from the top and one side of the ring, we could discern the orientation of the C-H bonds in relation to the C-C ring system. Let's construct the molecule piece by piece and see what it would look like.

❖ First, we draw two parallel lines. These lines are two of the six C-C bonds of cyclohexane.

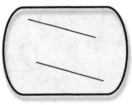

❖ Next, we connect the two left endpoints of the lines to two additional lines (two more C-C bonds) that converge to a single point below the parallel lines:

❖ Finally, we connect the right endpoints of the parallel lines to two additional lines (the last two C-C bonds) that converge to a single point above the parallel lines:

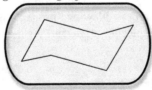

This completes the carbon framework viewed from slightly overhead and to one side. The drawing we have just completed looks a little like a chaise lounge and is referred to as the "chair" conformation of cyclohexane. Now, to complete the structure, let's draw in the C-H bonds.

❖ We begin by drawing six lines (C-H bonds) that are parallel to alternate C-C bonds and placing a hydrogen atom on the end of each line:

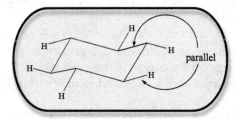

Notice that the C-H bonds are parallel with the C-C bonds one carbon removed from them. These C-H bonds are oriented around the equator if you will; they are called equatorial C-H bonds.

But cyclohexane has twelve hydrogens, so we must add the remaining six to our structure – one more to each carbon.

❖ Beginning at the rightmost apex, we place a C-H bond up, and as we proceed around the ring, we alternate. The next C-H bond goes down and so on until all six are installed:

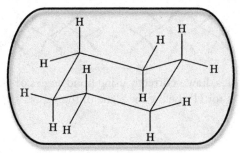

The six C-H bonds that are oriented vertically are referred to as axial C-H bonds.

Cyclohexane adopts the chair conformation to achieve C-C and C-H bond angles of 109.5°, the preferred bond angle found in molecules with sp^3 hybridization and tetrahedral geometry. It turns out that this is the lowest energy conformation cyclohexane can adopt. Although there are other conformations, we'll discuss those and why they are not as energetically favorable in Chapter 4.

Problem 3.5. Construct a model of the chair form of cyclohexane with the (a) axial hydrogens and (b) the equatorial hydrogens.

3.5. Summary

Drawing chemical structures is the application of chemical principles that we learned in Chapter 2. From the Lewis electron dot structure, we get a visual flavor for keeping tab of all valence electrons in a molecule, radical or ionic species. We apply a little shorthand in the Kekulé structure, and still more in the skeletal structure, remembering they are Lewis electron dot structures in disguise. The Newman projection gives us insight into the interactions between atoms in molecules. Correctly drawn chemical structures must account for bonding and non-bonding electrons and geometric considerations. With knowledge of these concepts, we can now apply them systematically to more complex cases and chemical reactions.

Solved Problems

Problems **3.1**, **3.4** and **3.5** require construction of models; no solutions are provided.

3.2.
Given: (a) 2-pentanol and (b) the anion of 2-pentanol, 2-pentoxide.
Find: Calculate the formal charge on the oxygen in these species.
Plan: Use formula: F.C. = # valence shell e⁻ of neutral atom – ½ # of bonding e⁻ – # of nonbonding e⁻
Solve: (a) 2-pentanol: O: $6 – ½(4) – 4 = 0$, (b) 2-pentoxide: O: $6 – ½(2) – 6 = -1$.
Check: Formal charges on oxygen calculated correctly.

3.3.
Given: skeletal structure of 2-pentanol.
Find: Expand skeletal structure to a sawhorse projection that depicts all hydrogens.
Plan: Draw the structure of 2-pentanol using tetrahedral bond angles, wedges for H's above the plane of the paper and dashs for H's below the plane of the paper.
Solve:

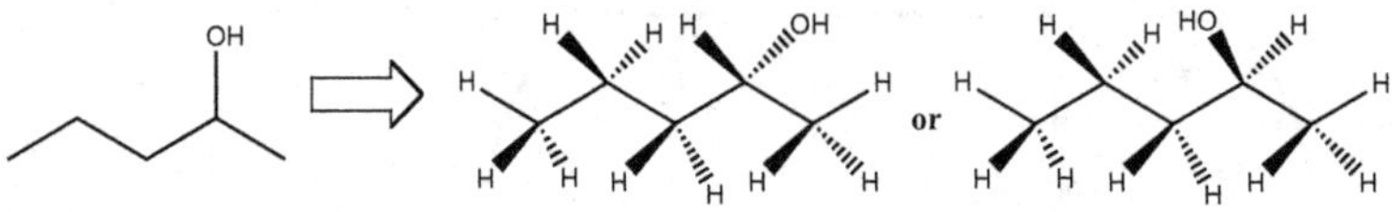

skeletal structure   sawhorse projection

Check: Structure of 2-pentanol is drawn correctly using bond angles of 109°, wedges for H's above the plane of the paper and dashs for H's below the plane of the paper.

Chapter 4 – Stability of Carbon Compounds and Reactive Intermediates

One major factor in determining the manner and ease with which organic compounds react is their stability. Alkanes, for instance, adopt conformations to achieve the lowest energy possible. Stability of alkenes and aromatic compounds similarly depend on steric and electronic factors.

Reactive intermediates are species that arise temporarily as a result of some chemical transformation along a reaction pathway between reactants and products. These species usually are higher in energy than either the reactants or products and most have limited lifetimes. Their higher energy causes them to react readily with other molecules to yield final products which have a lower energy.

In many cases, knowing relative stability trends within similar types of molecular species can allow us to predict how molecules will react as well as determine what products may arise from reactions. Although an exhaustive study of the types of compounds and intermediates found in organic chemistry is beyond the scope of this text, we will examine a number of cases the introductory organic student can expect to encounter. We will first address the stability of acyclic and cyclic alkanes by conformational analysis. Next, we will investigate alkene stability and the unusual stability of aromatic compounds. Following that, we will examine several reactive intermediates: carbocations, carbon radicals, and carbanions. First, let's look at alkanes.

4.1 Alkane Stability

Conformational analysis is a key method for examining the stability of organic molecules. This concept will help us predict relative stability of molecules from the alkane family.

4.1.a. Acyclic (non-cyclic) Alkanes

Recall from chapter 3 that we may draw a Newman projection of an organic molecule (Figure 3.10) from a given Kekulé, line-bond or skeletal structure. We will now examine the three types of Newman projections using the molecule butane as an example.

➢ Staggered Conformations of Butane

Butane, C_4H_{10}, may be represented in a Newman projection by taking the three-dimensional line-bond structure (sawhorse projection) and sighting down one of the C-C bond axes. For our example, we will sight down the C2-C3 axis:

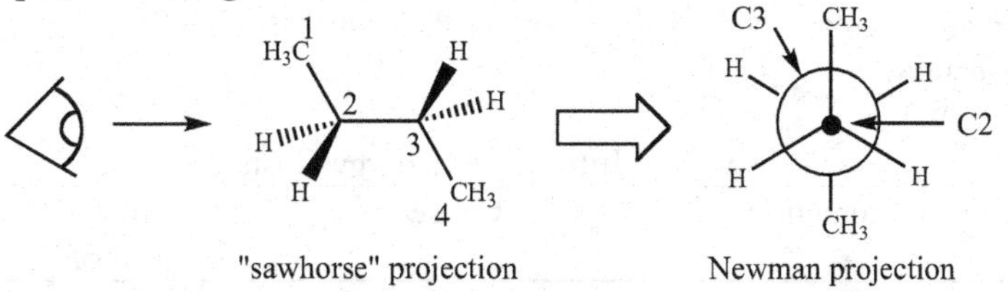

Figure 4.1. Staggered Newman Projection of Butane.

As we can see, not only are the hydrogens spaced as far apart as possible, but the large methyl groups are on opposite sides of the projection. Therefore, we say this is the **anti** conformation and as such, is the lowest energy conformation possible since no atoms on adjacent carbons are close enough to interact significantly.

If we were to rotate the number 3 carbon of butane 120°, however, we would encounter another conformation, the "**gauche**" conformation, shown in Figure 4.2. In our example, we will draw in the C-H bonds for the methyl groups:

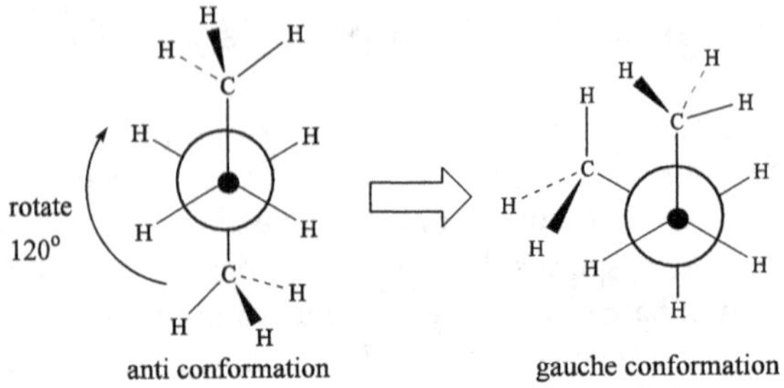

Figure 4.2. Gauche Newman Projection of Butane.

This rotational isomer or "rotamer" is known as the "gauche" conformation because now the carbon atoms of the methyl groups are close to one another. This closeness produces steric strain on the molecule, resulting in the gauche conformation being about 3.8 kJ/mol higher in energy than the anti conformation.

> Eclipsed Conformations of Butane

Although the gauche conformation results in some interaction between the methyl groups, if we rotate the number 3 carbon an additional 60° clockwise, we obtain a conformation where all the atoms attached to the C2 and C3 carbons are in line with each other – an **eclipsed** conformation.

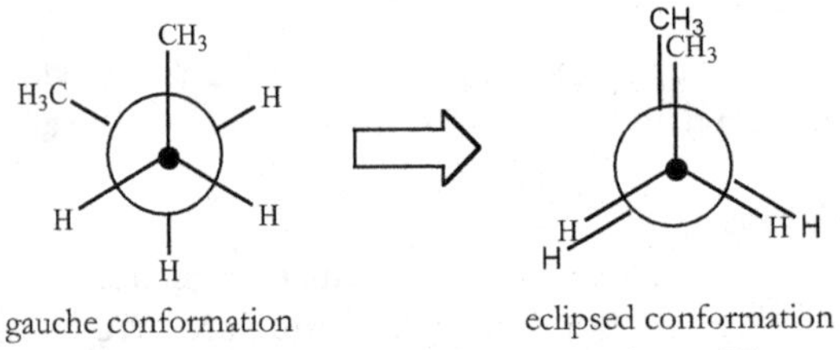

Figure 4.3. Eclipsed Newman Projection of Butane.

It turns out that this conformation is the highest in energy of all the conformations of butane, having a total of three eclipsing interactions: 2 H↔H, and 1 CH_3↔CH_3. These H↔H interactions cause the molecule to twist and produce torsion, while the CH_3↔CH_3 interactions produce both steric and torsional strain - that associated with the large groups being in close proximity to each other. Summarized below are calculated values for interactions leading to higher energy conformations of acyclic organic molecules.

Table 4.1. Interaction Energy Costs.[1]

Interaction	Cause	Energy Cost (kJ/mol)
CH_3↔CH_3 gauche	steric strain	3.8
H↔H eclipsed	torsional strain	4.0
H↔CH_3 eclipsed	torsional strain	6.0
CH_3↔CH_3 eclipsed	torsional + steric strain	11

As can be seen from the table, the eclipsed conformation of butane shown above has an energy cost of 2(4.0kJ/mol) + 11kJ/mol = 19kJ/mol. This means that the anti staggered conformation of butane (Figure 4.1) is 19kJ/mol more stable than the eclipsed form in Figure 4.3.

[1]Kingsbury, C. *J. Chem. Ed.*, **1979**, *56*, 431; Wiberg, K., et al. *J. Am. Chem. Soc.*, **1988**, *110*, 8029; Allinger, N., et al. *J. Am. Chem. Soc.*, **1990**, *112*, 114.

Of course, three other conformations of butane exist, one staggered and two eclipsed. Each of these would have different energies, according to the interactions present in the conformations.

Example Problem 1. Sighting down the C2-C3 axis, calculate the strain energy associated with the gauche conformation of 2,3-dimethylbutane shown below:

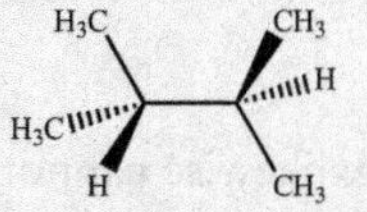

Given: Sawhorse projection of 2,3-dimethylbutane shown above.
Find: Steric strain energy of this gauche conformation.
Plan: (1) Redraw structure as a Newman projection.
　　　(2) Determine the gauche interaction(s).
　　　(3) Use Table 4.1 to obtain the energy cost and add all interactions together for the total energy cost.
Solve:

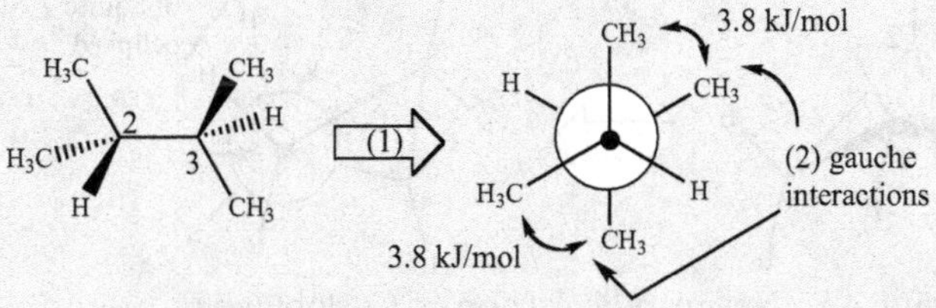

(3) 2 gauche interactions x 3.8 kJ/mol = 7.6 kJ/mol

Check: (1) Newman projection is drawn correctly; (2) two gauche interactions were found; (3) Total energy cost is calculated correctly.

Problem 4.1. Draw the other Newman projections for the butane molecule (Figure 4.1) and calculate their energies.

Problem 4.2. For 2-methylbutane, sighting down the C2-C3 axis:
a. Draw a Newman projection of the most stable conformation.
b. Draw a Newman projection of the least stable conformation.

4.1.b. Cyclic Alkanes

　　Cycloalkanes likewise have conformations that differ in energy according to interactions between substituents. In addition to torsional and steric strain found in acyclic alkanes, cycloalkanes also experience ring strain in cases where the ring system cannot adopt the preferred tetrahedral arrangement of atoms and achieve bond angles of 109.5°. Let's begin with the simplest ring system, cyclopropane.

➤ Conformation of Cyclopropane

　　The three membered ring, cyclopropane, is very highly strained since its bond angles, as dictated by geometry, are about 60° and not 109.5° as would be ideally expected for sp³ hybridized carbons. Despite this discrepancy, cyclopropanes are found in natural products and can be synthesized in the laboratory. Shown below in Figure 4.4 is the only Newman projection of cyclopropane. Sighting down the C1-C2 bond, we see all the C-H bonds are eclipsed.

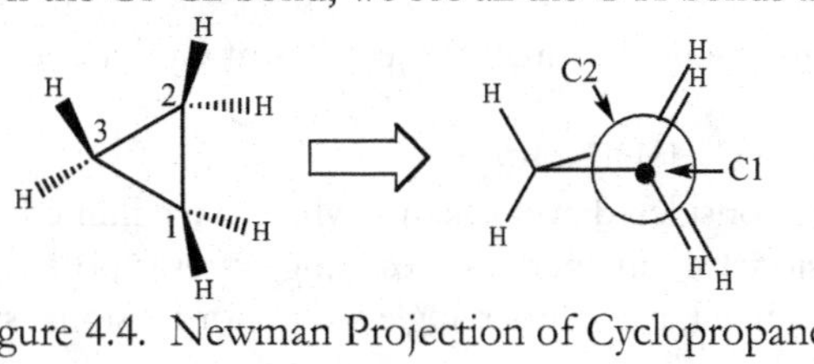

Figure 4.4. Newman Projection of Cyclopropane.

33

To offset some of the torsional strain from the eclipsing hydrogens and the ring strain associated with such a severe bond angle requirement, cyclopropane's molecular orbitals adopt a different geometry. As opposed to the extensive head-on sp³-sp³ overlap normally found in the σ bonds of most alkanes, the C-C bonds of cyclopropane are bent slightly (sometimes called banana bonds). This weakens the σ bonds of cyclopropane, making it more reactive (less stable).

➤ Conformation of Cyclobutane

Cyclobutane, a four-membered ring, likewise cannot achieve the desired 109.5° bond angles associated with the tetrahedral geometry of acyclic alkanes. The bond angles for its carbon-carbon framework only reach 90°, causing angle strain which, while less than that of cyclopropane, is substantial. However, the torsional strain associated with its large number of hydrogens causes cyclobutane to adopt a "puckered" or bent geometry. While this raises the angle strain energy, it reduces the torsional strain in the molecule by relieving the eclipsing H↔H interactions that occur in cyclopropane. This minimum energy conformation is shown in figure 4.5.

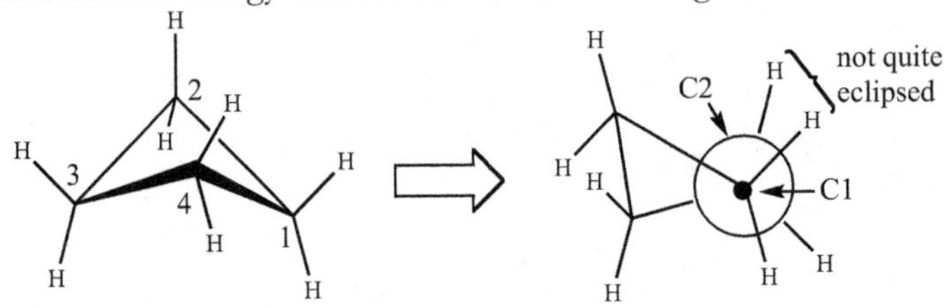

Figure 4.5. Newman Projection of Cyclobutane.

➤ Conformations of Cyclopentane

If cyclopentane, a five-membered ring, were planar, it would have very little angle strain, achieving the desired 109.5° bond angles. Yet, cyclopentane twists or "puckers" to adopt a geometry to minimize the substantial torsional strain its large number of hydrogens causes. While the puckering raises the angle strain energy, two conformations that are close in overall energy result in which all the hydrogens are nearly staggered. These conformations, the **envelope** and **half-chair**, are shown in Figure 4.6.

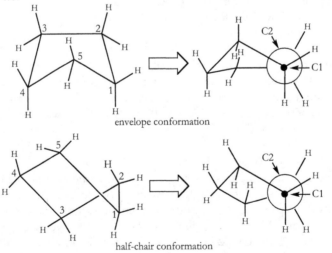

envelope conformation

half-chair conformation

Figure 4.6. Newman Projections of Cyclopentane.

➤ Conformations of Cyclohexane

Up to now, we have considered cycloalkanes which have limited conformational mobility because of the ring and torsional strain energies these ring systems possess. Cyclohexane, a six-membered ring, on the other hand, has neither angle strain nor torsional strain. It achieves the desired 109.5° bond angles by adopting a **chair** conformation (introduced in Chapter 3). Recall that the hydrogens oriented vertically on the ring are called axial hydrogens, while those projected from the periphery of the ring are called equatorial hydrogens. To understand why this conformation is preferred, let's look at a Newman projection of cyclohexane. In Figure 4.7, we sight down the C1-C2 and C5-C4 axes for our perspective.

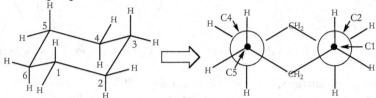

Figure 4.7. Newman Projection of Cyclohexane.

As can be seen from the Newman projection above, all the hydrogens are staggered so there is no torsional strain in the molecule.

Cyclohexane also exhibits conformational mobility – that is, we can have interconversion to another conformation. The first of these interconversions is chair to chair. This is accomplished by a "**ring-flip**". To effect this conformational change, the number 6 carbon moves from a position below the plane of the ring defined by the C1-C2 and C5-C4 bonds to one above the ring plane. Meanwhile the number 3 carbon moves from a position above the plane of the ring to one below the ring plane. See Figure 4.8.

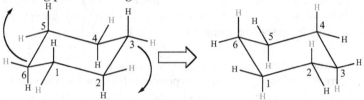

Figure 4.8. Ring-Flip of Cyclohexane.

An important consequence arises from this interconversion. Notice the positions of the equatorial hydrogens indicated in gray after ring-flip is complete. They are now axial. The axial hydrogens, upon ring-flip, are now equatorial. Notice also that **ring-flip occurs by rotation about single bonds and not bond cleavage**.

Another conformation of cyclohexane called the boat is depicted in Figure 4.9. Although the boat conformation has no angle strain, several other factors make it less stable than the chair conformation. Let's see why this is the case.

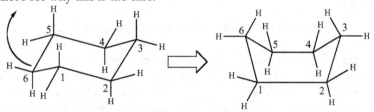

Figure 4.9. Boat Conformation of Cyclohexane.

By interconversion to the boat conformation, we introduce steric strain between the hydrogens on C3 and C6 where the two C-H bonds are pointing toward each other. There is also torsional strain arising from the H↔H eclipsing interactions, which we can depict by showing the Newman projection in Figure 4.10:

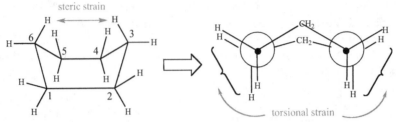

Figure 4.10. Newman Projection of the Boat Conformation of Cyclohexane.

As a result of these strain energies, boat cyclohexane is about 29 kJ/mol less stable than chair cyclohexane. By twisting, the boat conformation can reduce its energy by approximately 5.5 kJ/mol, but this is still a much more strained conformation than the chair.

Why would the cyclohexane ever adopt the boat conformation, given the stability of the chair conformation? The answer is that for some cyclohexane derivatives, reaction pathways require temporary adoption of a boat conformation on the way to forming product. There are also certain cycloalkane substituents that interact most favorably in a boat conformation. Those cases, while rare, do occur.

4.1.c. Conformational Analysis of Substituted Cyclohexanes

Now that we have a basic understanding of the various conformations of cycloalkanes, we need to address further the stability of substituted cyclohexanes. We know that steric factors can play a role in the stability of alkane conformations. We will begin by examining the effect the methyl group has on the conformational stability of methylcyclohexane.

➢ Monosubstituted Cyclohexanes – Methylcyclohexane

Consider Figure 4.11 which shows the two possible chair conformations of methylcyclohexane – one where the methyl group is axial and one where the methyl group is equatorial. In the conformation on the left, the axial methyl group has what are known as 1,3-diaxial interactions with the two axial hydrogens due to their proximity. These interactions increase the energy of this conformation, and cause a shift in the equilibrium to the right, toward the conformation with the methyl group in an equatorial orientation.

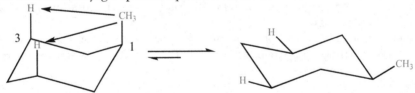

Figure 4.11. Methylcyclohexane Conformations.

It turns out that this is the case for most monosubstituted cyclohexanes; bulky functional groups prefer equatorial orientations when possible to avoid 1,3-diaxial interactions that arise in the other conformation. How much energy difference are we talking about? The table below gives some figures, which show an increase in 1,3-diaxial strain energy as the size of the substituent increases.

Table 4.2. Monosubstituted Cyclohexane Strain Energies.[2]

Substituent (X)	Strain Energy due to one H-X 1,3-diaxial interaction (kJ/mol)
-CN	0.35
-Cl	0.90
-OH	1.82
$-CH_3$	3.64
$-C_2H_5$	3.75
$-CH(CH_3)_2$	4.62
$-C(CH_3)_3$	>11.3

➤ Disubstituted Cyclohexanes – *t*-Butyl-methylcyclohexanes

There are three cases we will now examine to illustrate the stability of cyclohexanes that are substituted in two positions. The *t*-butyl group is very large and dictates the conformational equilibrium of cyclohexanes which bear it as a substituent. From this discussion, you will be able to predict the most stable conformation a disubstituted cyclohexane species is likely to adopt.

In the cases that follow, we introduce a new concept, that of cis and trans isomerism in cycloalkanes. While we will cover this in chapter 5, the cis and trans isomers refer to the spatial arrangement of the methyl and *t*-butyl groups with respect to the plane defined by the C2-C3 and C5-C6 bonds of the cyclohexane ring.

✓ 1-*t*-Butyl-2-methylcyclohexane Conformations

Consider the chair conformations of trans- and cis-1-*t*-butyl-2-methylcyclohexane shown below in Figure 4.12. In this figure, as well as in those that follow, we depict only the axial hydrogens in an effort to simplify our conformational stability study.

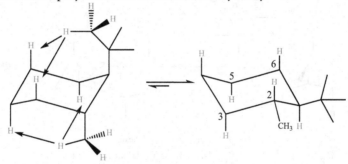

trans-1-*t*-butyl-2-methylcyclohexane

cis-1-*t*-butyl-2-methylcyclohexane

Figure 4.12. 1-*t*-Butyl-2-methylcyclohexane Conformations.

[2]Carroll, F. A. Perspectives on Structure and Mechanism in Organic Chemistry. Brooks/Cole Publishing : Pacific Grove, CA, 1998, p. 138.

In the trans-1,2 case, the left conformation places both alkyl substituents axial; this is energetically unfavorable since it maximizes the 1,3-diaxial interactions. The conformation on the right eliminates these interactions completely. In fact, the equilibrium constant, $K_{axial \rightarrow equatorial}$, is greater than 100.

Even though the cis-1,2 conformation always has a 1,3-diaxial interaction and a $CH_3 \leftrightarrow C(CH_3)_3$ gauche interaction, the conformation on the right is more stable because the $H \leftrightarrow CH_3$ interactions are much smaller than the $H \leftrightarrow C(CH_3)_3$ 1,3-diaxial interactions. **Note going from axial to equatorial requires bond rotation, but going from cis to trans requires bond cleavage.**

    ✓   1-*t*-Butyl-3-methylcyclohexane Conformations

Now, let's look at the chair conformations of trans- and cis-1-*t*-butyl-3-methylcyclohexane shown below.

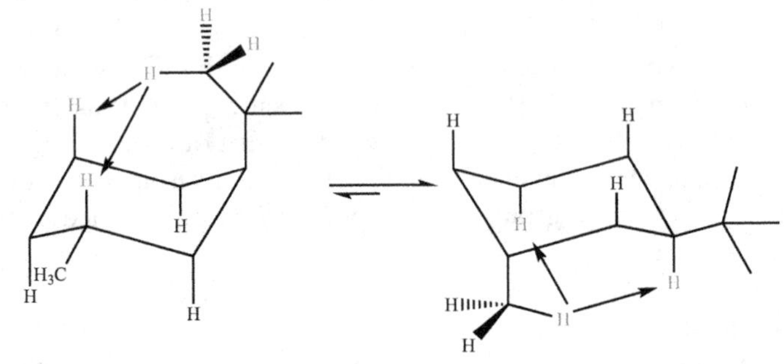

trans-1-*t*-butyl-3-methylcyclohexane

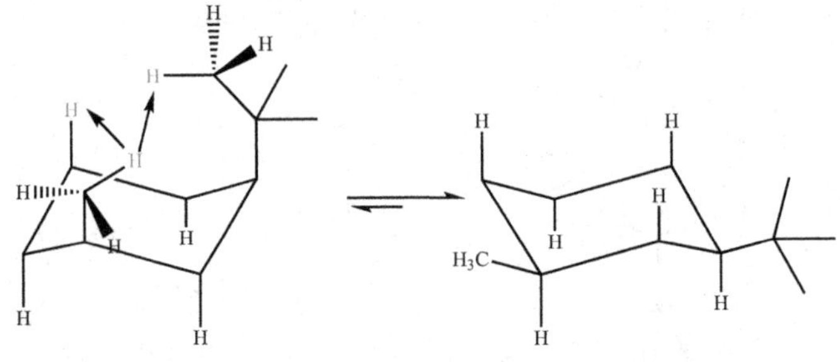

cis-1-*t*-butyl-3-methylcyclohexane

Figure 4.13. 1-*t*-Butyl-3-methylcyclohexane Conformations.

Here, the trans-1,3 isomer will always have one group that is subject to 1,3-diaxial interactions; the right conformation is more stable for the same reasons as cited above for the cis-1,2 case. Whereas, the cis-1,3 isomer can ameliorate the 1,3-diaxial interactions by ring-flip to the conformation where both alkyl groups are equatorial.

    ✓   1-*t*-Butyl-4-methylcyclohexane Conformations

Consider the chair conformations of 1-*t*-butyl-4-methylcyclohexane shown in Figure 4.14.

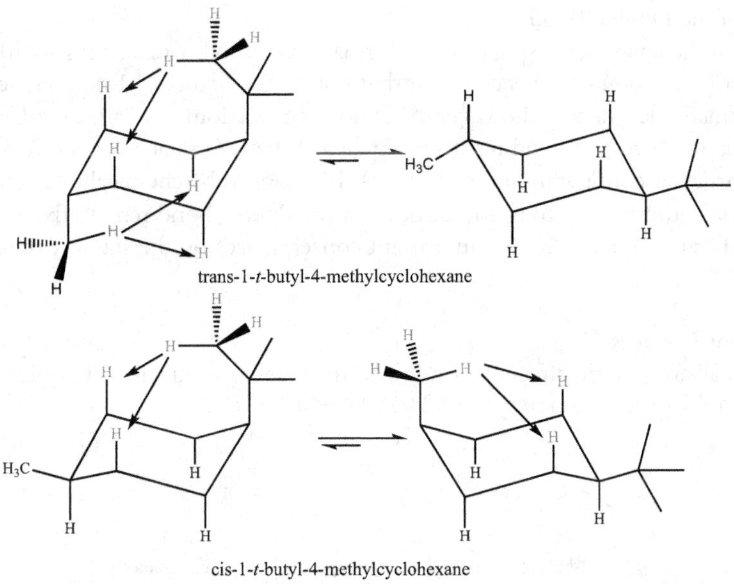

trans-1-*t*-butyl-4-methylcyclohexane

cis-1-*t*-butyl-4-methylcyclohexane

Figure 4.14. 1-*t*-Butyl-4-methylcyclohexane Conformations.

The cis-1,4 isomer will always contain a pair of 1,3-diaxial interactions, but the right conformation will be more stable since the larger *t*-butyl group will occupy the more spacious equatorial position. In the trans-1,4 isomer, the equilibrium lies far to the right to avoid the numerous H↔C(CH$_3$)$_3$ 1,3-diaxial interactions. The data for these isomeric *t*-butyl-methylcyclohexanes is summarized below. This table may also be applied to any cyclohexane substituted with two alkyl groups of differing sizes where the large group occupies the 1-position.

Table 4.3. Conformational Stability of Disubstituted Cyclohexanes.

Isomer	Most stable	Least stable	Remarks
trans-1,2	e,e	a,a	
cis-1,2	e*,a	a,e	*large group equatorial
trans-1,3	e*,a	a,e	*large group equatorial
cis-1,3	e,e	a,a	
trans-1,4	e,e	a,a	
cis-1,4	e*,a	a,e	*large group equatorial

e=equatorial, a=axial

In summary, when cyclohexanes contain one or two alkyl substituents, the conformation that provides either for the largest group or both groups to be in an equatorial orientation will normally be the most stable. Finally, the conformational equilibrium will lie in the direction of the most stable conformation.

4.2 Alkene Stability

As we learned in Chapter 2, alkenes are organic molecules which contain a C=C double bond. Double bonds, you will recall, are composed of one σ bond and one π bond. For alkenes, this means that two sp^2 hybrid orbitals overlap in a head-on fashion to produce the σ bond, and the p orbitals on those carbons overlap sideways to produce the π bond. The stability of alkenes is best understood by examining the steric and electronic consequences of the bonding that makes them up.

4.2.a. Rigidity of the Double Bond

Earlier in the chapter, we spoke of conformational mobility in alkanes – carbon atoms bonded together by a σ bond can rotate about that bond. See Figures 4.1-4.3. Hence, we can have different conformations. Alkenes, however, have no such freedom to change conformations. The p-p orbital overlap to form a π bond results in rigidity of the C-C skeletal framework, locking the molecule in a particular configuration that may only be changed by chemically breaking the π bond.

Therefore, atoms attached to the carbons of the double bond remain fixed on one side of the double bond or the other. This has important consequences on the stability of alkenes as we shall see.

4.2.b. Steric Consequences

As with alkanes, steric effects play a role in the stability of alkenes. Consider the following alkenes, in which the order of stability is 1<2<3<4<5<6:

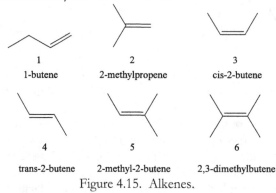

Figure 4.15. Alkenes.

Compound 1 has no steric strain. In compound 2, the methyl groups attached to C2 are in close proximity to one another and produce steric strain. In compound 3, the methyl groups are now further apart, which lessens the steric strain slightly. Compound 4 has even lower steric strain because the methyl groups are on opposite sides of the double bond. Compounds 5 and 6, however cannot avoid the steric strain associated with the methyl groups' proximity, so why is it that they are more stable than compounds 1-4? Factors other than steric strain must be more important.

4.2.c. Hybridization Consequences

In chapter 2, we discussed hybridization and as we learned, alkene carbons are sp² hybridized. This fact explains the planar geometry of the molecules in the above figure. In addition, sp² hybridization has important consequences on bond strengths of C-C bonds that helps explain why replacement of hydrogens with other groups (increased substitution) around the double bond makes it more stable. Carbon's sp² hybrid orbitals have more s character (33%) than the sp³ hybrid orbitals (25%) found in alkanes. As a result, the sp² hybrids are less directional, and are slightly larger in volume. This permits better overlap with other orbitals, shorter bond length and hence, higher bond strength. Consider compounds 1, 2, 3 and 4. Compound 1 contains one =C-CH₂ (sp²-sp³) bond and one -CH₂-CH₃ (sp³-sp³) bond while compounds 2, 3 and 4 contain two =C-CH₃ (sp²-sp³) bonds. The additional bond strength derived from these bonds makes compounds 2, 3 and 4 more stable than compound 1. However, another factor, hyperconjugation also plays a critical role in the stability of alkenes.

4.2.d. Hyperconjugation Consequences

Alkyl groups attached to the double bond of an alkene are able to stabilize the alkene through a phenomenon known as hyperconjugation. Recall the hydrogen story from Chapter 2, where we found that a two-electron, two-MO interaction was stabilizing. In alkenes, alkyl C-H sp³ filled hybrid orbitals can align with the unfilled antibonding $\pi *$ orbital of the alkene shown in Figure 4.16, giving rise to a two-electron, two-MO interaction.

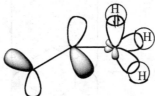

Figure 4.16. Interaction Between an Alkene $\pi *$ Orbital and a C-H sp³ Hybrid Orbital.

The electron density in adjacent C-H bonds of neighboring methyl groups stabilizes the alkene by hyperconjugation with the $\pi *$ orbital. The more alkyl C-H groups available for this type of interaction, the more stable the alkene. Hence, 2,3-dimethylbutene is the most stable of the alkenes in Figure 4.15.

4.3. Stability of Aromatic Compounds

Early in the history of organic chemistry, chemists had few means of distinguishing compounds from one another. A common test to verify that one had prepared a compound was to smell its aroma or odor. For many molecules, their aroma was characteristic. For example, the aroma of cinnamon is easily recognizable during baking. The responsible organic compound for this odor is cinnamaldehyde, which contains a benzene ring. Indeed, many molecules which possess familiar smells have a benzene ring. Thus, the term aromatic became synonymous with compounds containing benzene rings. Some examples are below.

trans-cinnamaldehyde   benzaldehyde   vanillin
(cinnamon)   (almond oil)   (vanilla bean)

Aromatic compounds, as a group, are probably the most studied of all organic molecule classes. Their presence in dyes, medicines, fragrances and complex biomolecules makes them the focus of numerous investigations. Why are they ubiquitous in nature? What is it about the benzene ring that makes it so special? The answers to these questions center around benzene's and other aromatic molecules' stability.

4.3.a. The Uniqueness of Benzene

After the discovery of benzene, chemists were puzzled for years at its lack of reactivity. Reagents that reacted with molecules containing multiple bonds did not react with benzene. Scientists eventually probed the unusual stability of benzene by comparing heats of hydrogenation ($\Delta H°_{hydrog}$), that is, the heat given off by the addition of H_2 gas to the series of cyclohexenes below to yield cyclohexane. What they found was astonishing. As expected, the value for 1,3-cyclohexadiene was nearly twice that of cyclohexene. It was reasoned that if benzene, "1,3,5-cyclohexatriene", were hydrogenated, a value nearly three times that of cyclohexene would be obtained. Not only was this not found to be the case, but a $\Delta H°_{hydrog}$ value less than that for 1,3-cyclohexadiene was observed. See Figure 4.17.

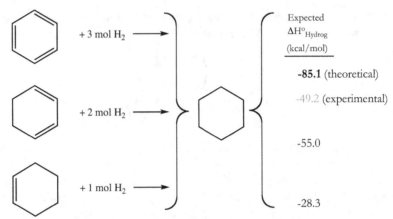

		Expected ΔH°_{Hydrog} (kcal/mol)
+ 3 mol H_2 →		**-85.1** (theoretical)
		-49.2 (experimental)
+ 2 mol H_2 →		-55.0
+ 1 mol H_2 →		-28.3

Figure 4.17. Cyclohexene Heats of Hydrogenation.

This difference was difficult to explain based on the theories available at the time. Furthermore, benzene has some other unusual characteristics. One of them is the length of the C-C bonds – they are all the same. What's more, the bond length is in between those of typical C=C bonds (1.34Å) and C-C bonds (1.54Å).

This finding is inconsistent with bond lengths found in molecules containing both double and single bonds. Another electronic factor, resonance, or electron delocalization, will help us to better understand these experimental observations.

4.3.b. The Resonance Energy of Benzene

The difference in the expected and actual ΔH°_{hydrog} values for "1,3,5-cyclohexatriene" and the aforementioned properties of benzene can be explained by resonance theory. Resonance is a phenomenon that allows us to describe the placement of π and nonbonding electrons in the Kekulé structure of a molecule in more than one way. We may move these electrons without altering the atomic arrangements. For benzene, we have two resonance forms. These resonance forms are not what the actual molecule looks like; rather the molecule contains characteristics of both forms and so is a composite, or hybrid, of the resonance forms. The π system is denoted by a circle and thus, does not specifically indicate the number of π electrons, An alternative molecular orbital picture, shown on the far right, is descriptive of the special nature of benzene, where the six p orbitals that make up the entire π system is a representation that more accurately describes the equivalent nature of the C-C bonding in benzene.

resonance forms   resonance   molecular orbital
                         hybrid   picture of benzene

We typically show the movement of electrons by means of "electron-pushing" arrows from an electron-rich center (in this case, a double bond) toward an electron-poor center. We indicate the

42

different resonance forms by a double-headed arrow between them. The extra stability of benzene relative to the hypothetical "1,3,5-cyclohexatriene" (35.9 kcal/mol) has been termed the resonance energy of benzene. In addition, the possibility of resonance helps explain why benzene's carbon-carbon bond lengths are the same. In support of this, studies of the electrostatic potential mapping in benzene show that the electron density throughout the ring is uniform. Since this is the case, it makes sense that the C-C bond lengths would be the same.

4.4 Carbocation Stability

4.4.a. Simple Carbocations

A carbon atom becomes electron deficient when its electron density is somehow depleted. In Chapter 2, we discussed the concepts of inductive bond polarization by electronegativity differences between atoms. If we remove a single electron from a neutral carbon atom, a carbon bearing a full positive (+1) charge is formed, a **carbon cation** or carbocation. Consider Figure 4.18 which shows 2-bromo-2-methylpropane reacting with chloride to produce 2-chloro-2-methylpropane and bromide:

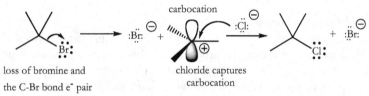

loss of bromine and
the C-Br bond e⁻ pair

chloride captures
carbocation

Figure 4.18. Reaction of 2-Bromo-2-Methylpropane with Chloride.

In this nucleophilic substitution reaction (a reaction type we will cover in chapter 7), the carbocation species shown above is an intermediate that is generated upon departure of the bromine with the C-Br bonding electron pair. Capture of the carbocation by the chloride leads to the product, 2-chloro-2-methylpropane. Noteworthy is the planar geometry of this intermediate, arising from the sp^2 hybridization of the central carbon atom. We'll examine this mechanism more closely in Chapter 7.

But why should this reactive intermediate form in the first place? The answer is stability. This carbocation has a positively charged carbon surrounded by three methyl groups conferring on the ion sufficient stability to exist for a finite period of time such that a nucleophile like the chloride may capture it. Figure 4.19 shows how the empty p orbital of the carbocation is stabilized by the presence of electron density found in the neighboring methyl groups, a phenomenon known as hyperconjugation:

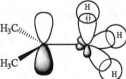

Figure 4.19. Tertiary (3°) Carbocation Stabilization by Hyperconjugation.

As we learned earlier in this chapter when we discussed alkene stabilization by hyperconjugation, the C-H sp^3 hybrid orbitals can align with the empty p orbital, providing electron density to stabilize the carbocation. Three methyl groups are available to assist in the stabilization. If fewer alkyl groups are present, less stabilization results. The stability trend for carbocations stabilized only by hyperconjugation is thus: 3° > 2° > 1° > methyl.

> Stabilization of 2° Carbocations by Rearrangement

Consider the 2° carbocations shown in Figure 4.20 which have either hydrogens or alkyl groups positioned on adjacent carbons.

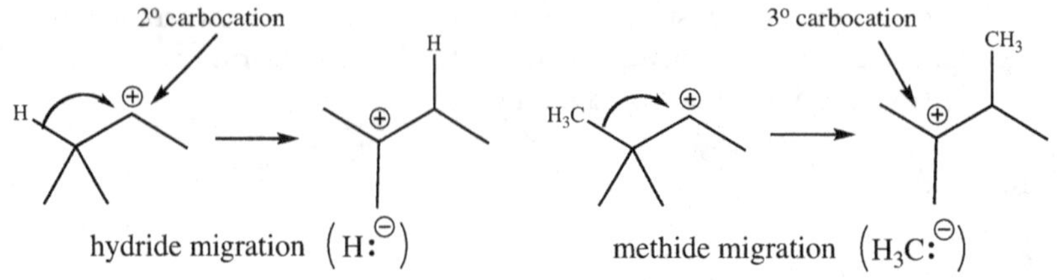

Figure 4.20. Carbocation Rearrangements.

While carbocations such as this are stabilized by hyperconjugation, migration of a hydrogen or alkyl group with its electron pair to the carbon with the empty p orbital produces a more stable 3° carbocation. Rearrangements like this (which we will discuss in Chapter 7) are commonplace during reactions of alkenes and can lead to formation of products in larger proportions than those resulting from normal carbocation capture.

4.4.b. Resonance Stabilized Carbocations

It turns out that many carbocations are not merely the simple cases illustrated above. Other factors may contribute more significantly to carbocation stability than hyperconjugation. For example, carbocations may be stabilized by donation of electron density from adjacent atoms with nonbonding electron pairs or through delocalization of charge through a π system.

> Stabilization of Carbocations by Electron Pair Donors

Consider the case shown in Figure 4.21, and predict its stability relative to that of the simple 3° carbocation depicted earlier.

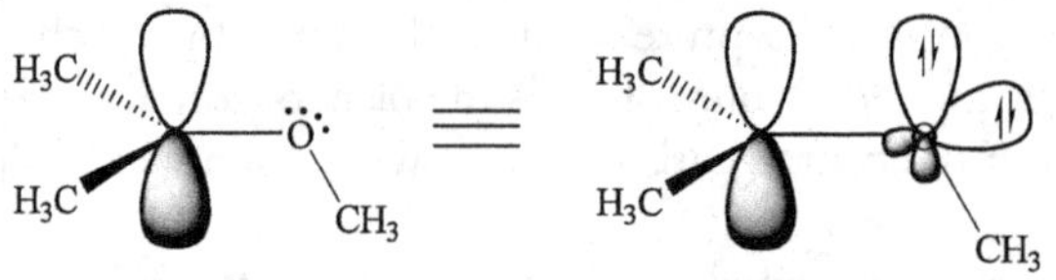

Figure 4.21. 2-Methoxypropyl Carbocation.

This tertiary carbocation can adopt a conformation where the sp³ hybrid orbitals containing the nonbonding electron pairs align for good overlap with the empty p orbital next door. This is a favorable two-electron, two-orbital interaction and the result is stabilization. Since these electrons are not contained in a bond, they may be shared with the p orbital in the form of a π bond. We can also express this using resonance forms that show the delocalization of the positive charge from the carbon to the oxygen:

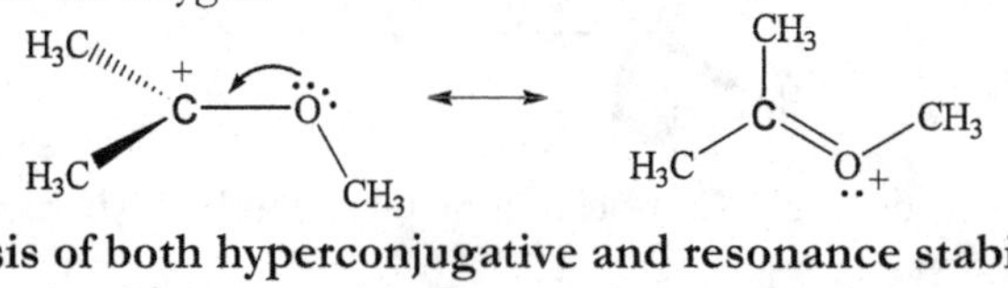

We predict, on the basis of both hyperconjugative and resonance stabilization that the 2-methoxypropyl carbocation is more stable than tertiary carbocations which are not stabilized by electron donating groups.

Problem 4.3. Like oxygen, sulfur can donate nonbonding electrons to stabilize empty p orbitals on adjacent carbons. (a) Draw resonance forms for $(CH_3)_2C^+(SCH_3)$, the 2-thiomethoxypropyl cation that show how delocalization of the positive charge takes place. (b) Compare this carbocation's stability to that of unstabilized 2° and 3° carbocations.

> ➢ Stabilization of Carbocations by π Systems

Pi bonds can also stabilize carbocations. Consider the case of the allylic carbocation shown in Figure 4.22.

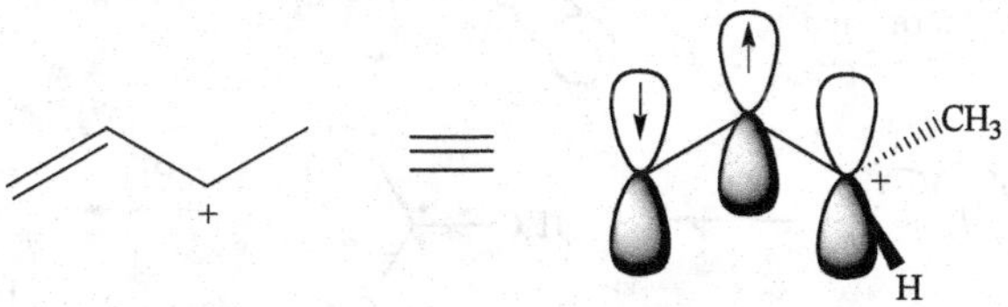

Figure 4.22. Allylic Carbocation.

Here, the π system is uninterrupted, which allows for delocalization of the two π electrons throughout the allylic carbocation molecular orbital. Depicted another way, we can show migration of the positive charge by resonance:

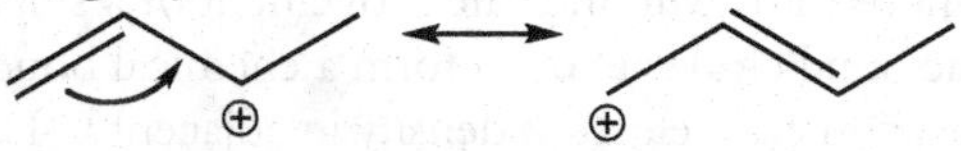

Before we leave this discussion, we should address the resonance forms shown above. Clearly the two **resonance forms** are not the same – the left form is a 2° allylic carbocation system with a monosubstituted alkene, while the right form is a 1° allylic carbocation with a disubstituted alkene. Applying our newfound understanding of hyperconjugation, we can predict the nature of this carbocation. The combination of stabilization effects by the adjacent π system and hyperconjugation make the 2° allylic carbocation the major contributor to the overall **resonance hybrid** of this allylic species shown below. Notice that a greater partial positive charge is present on the more highly substituted carbon:

Benzylic carbocations (Figure 4.23) are special cases in that they are extended allylic, cyclic systems. The stability of these cations is enhanced by the presence of the aromatic system adjacent to the electron deficient carbon center. Delocalization of the charge with the aromatic ring is possible:

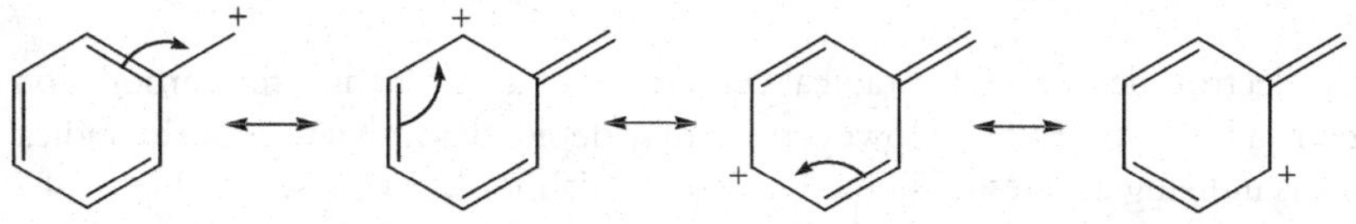

Figure 4.23. Resonance Forms of the Benzylic Carbocation.

Problem 4.4. As we learned earlier in this chapter, aromatic compounds owe their special stability to resonance. The benzylic cation depicted below is a 2° carbocation. Drawing resonance structures as needed, predict its stability relative to an unstabilized 2° carbocation. Compare the stability of this carbocation to a 3° carbocation.

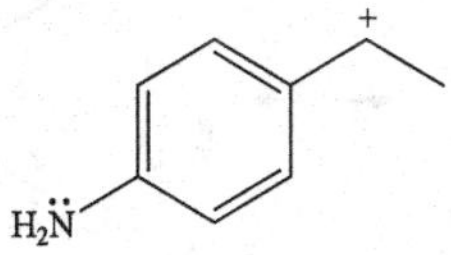

4.5. Stability of Carbon Radicals

Remember that any loss of electron density results in an electron deficient atom. Consider the formation of a tertiary carbon radical intermediate during the process of radical halogenation of 2-methylpropane in Figure 4.24:

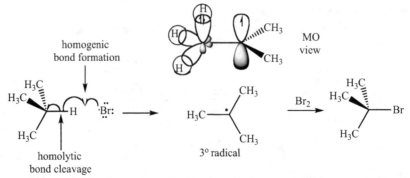

Figure 4.24. Formation of a Tertiary Carbon Radical Intermediate.

We use a single-headed or "fishhook" arrow to indicate movement of a single electron, either to form a radical species (homolytic bond cleavage) or to form a chemical bond (homogenic bond formation). Just as in the carbocation case, electron density in adjacent C-H bonds helps stabilize the radical center which is electron deficient with respect to the octet rule, although neutral. But, since a single electron is present in this p orbital instead of a full positive charge, hyperconjugative stabilization will have a smaller effect. Nevertheless, the radical stability trend is similar to that of carbocations: 3° > 2° > 1° > methyl.

Analogous arguments can be made for the stability of radicals with adjacent π systems, such as allylic and benzylic radicals shown in Figure 4.25. These double bonds serve to delocalize the radical and make the radical more stable.

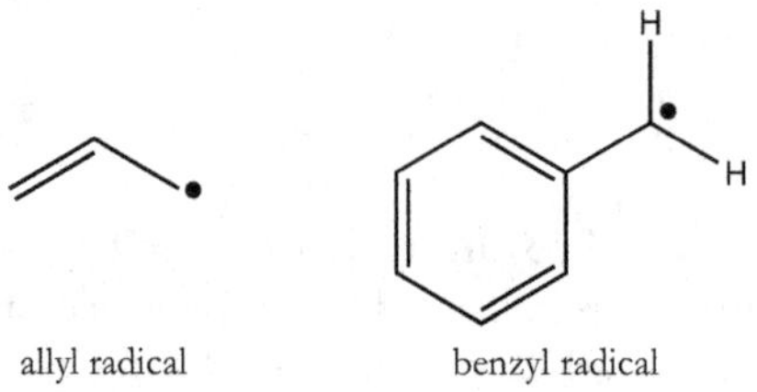

allyl radical   benzyl radical

Figure 4.25.   Allylic and Benzylic Radicals.

Of course, the electron deficiency for radical centers is not as marked as for carbocations, since there is a single electron in the p orbital. However, we may depict delocalization of the radical by drawing resonance forms utilizing a series of single-headed or "fishhook" arrows. See Figure 4.26.

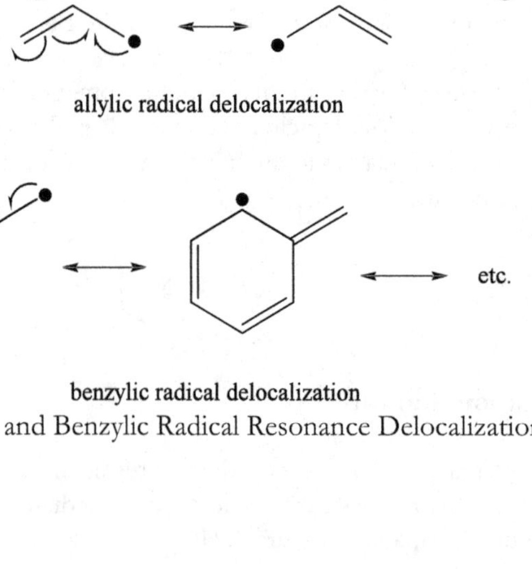

allylic radical delocalization

etc.

benzylic radical delocalization

Figure 4.26. Allylic and Benzylic Radical Resonance Delocalization.

Problem 4.5. The allylic and benzylic radicals depicted below are 2° radicals. Drawing resonance structures as needed, predict its stability relative to an unstabilized 2° alkyl radical. Compare the stability of these radicals to a 3° alkyl radical.

4.6. Stability of Carbanions

Carbon atoms containing an electron pair that is nonbonding in addition to three bonding pairs of electrons are called carbanions (**carbon anions**). Stability in anionic species has been widely studied, and many reactions in organic chemistry involve carbanion or carbanion-like intermediates. We begin our study of anions by considering the reaction of methyl bromide with lithium metal shown in Figure 4.27:

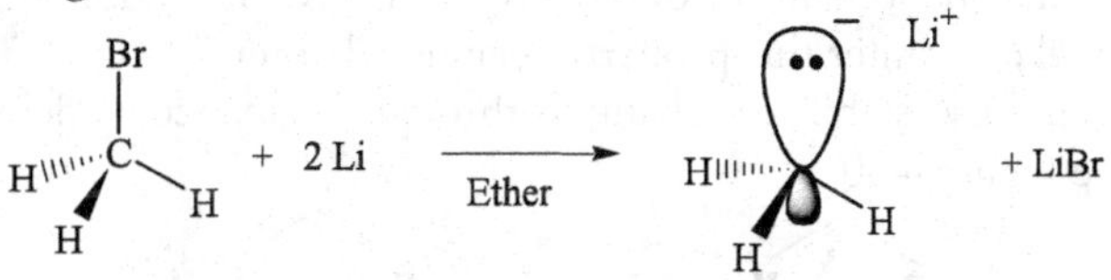

Figure 4.27. Methide Formation.

This reaction results in formation of an alkyllithium species, $R:^-Li^+$. The alkyl anion is extremely reactive and will function as a base or nucleophile, depending on reaction conditions.

What sorts of functional groups stabilize carbanions? Recall that alkyl groups stabilize carbocations by donation of electron density to the empty p orbital of the carbocation by hyperconjugation. Conversely, we would expect that the presence of adjacent alkyl groups or functional groups with nonbonding electron pairs would destabilize carbons with excess electron density. In the latter case, molecular orbital theory predicts destabilization since this would be a two-orbital, four-electron interaction.

Functional groups adjacent to the carbanion that bear a full or partial positive charge (electron withdrawing groups) as well as those that delocalize the anion through a π system would therefore be expected to stabilize a carbanion. The nitromethide ion shown in Figure 4.28 provides an example of carbanion stabilization by inductive electron withdrawal.

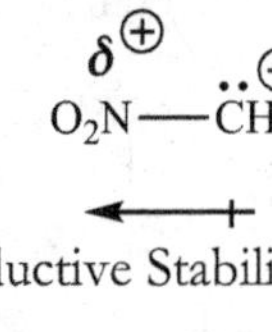

Figure 4.28. Inductive Stabilization of a Carbanion.

As shown in the figure above, the nitro (NO_2) functional group generates a δ^+ charge by inductive electron withdrawal (brought about through a difference in electronegativity between carbon and nitrogen). When electron withdrawing groups (EWGs) such as these are located adjacent to carbanions, the negative charge is stabilized.

Some EWGs can also stabilize carbanions by resonance, which, according to MO theory, is a favorable two-electron, two orbital interaction. In Figure 4.29 for example, resonance forms for the nitromethide anion, $^-:CH_2NO_2$, and the anion of acetone, $^-:CH_2COCH_3$, can be drawn that show the delocalization of the negative charge into the adjacent π system of the NO_2 and carbonyl groups:

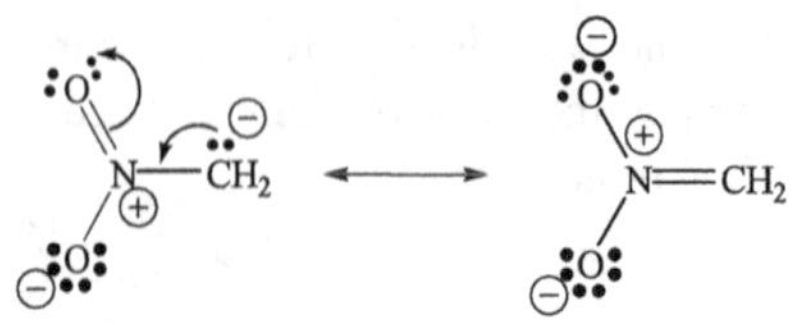

nitromethide anion

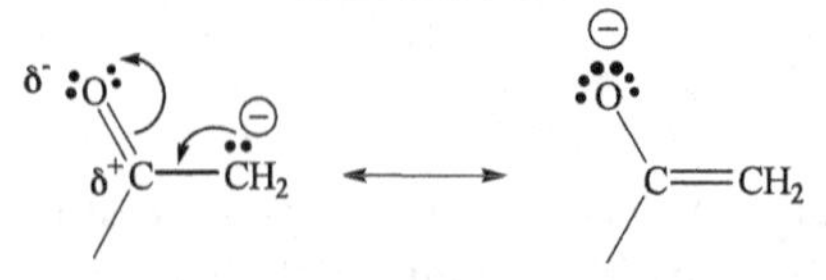

anion of acetone

Figure 4.29. EWG Resonance Stabilization of Carbanionic Species.

This illustrates an important phenomenon exhibited by some EWGs – additive inductive and resonance effects. We will encounter this property again in Chapter 7.

Adjacent π systems also stabilize carbanions through resonance by delocalizing the additional electrons. See Figure 4.30.

allylic carbanion delocalization

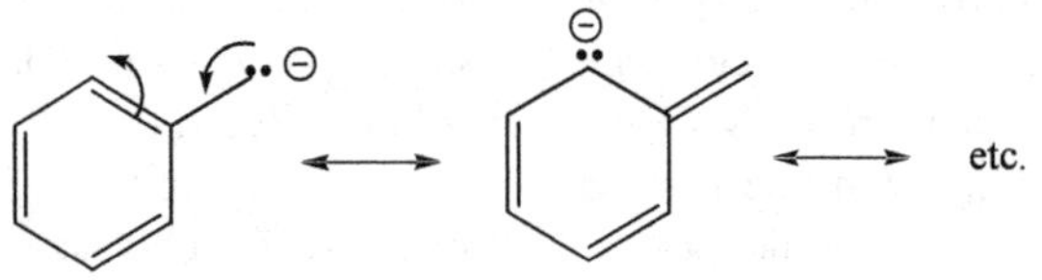

benzylic carbanion delocalization

Figure 4.30. Resonance Stabilization of Allylic and Benzylic Carbanions.

Problem 4.6. The Meisenheimer complex depicted below is formed upon nucleophilic addition to benzenes substituted with EWGs. Drawing resonance structures as needed, show how the π system of the aromatic ring and the EWG stabilize the carbanion.

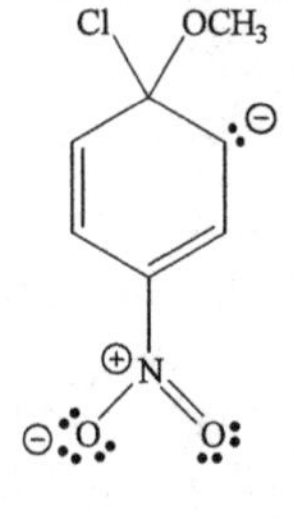

4.7. Summary

As we have seen, the stability of carbon species depends upon several factors: steric effects of nearby substituent groups, the ability of the species to adopt conformations or geometry that minimize strain energies, and the electronic surroundings. Stability in alkanes and cycloalkanes are functions of conformational flexibility – that is, adoption of lower energy conformations results in smaller steric effects. Alkene stability is also partially influenced by steric considerations. Species are also subject to electronic effects, which may arise from a combination of inductive and resonance contributions. These forces may act alone or in concert to enhance stability – hyperconjugative effects, resonance effects through adjacent π systems or inductive effects through adjacent σ systems.

48

Hyperconjugation is the term for stabilization of π systems by either adjacent bonding or nonbonding electron pairs. Alkyl groups provide this sort of interaction for alkenes and electron deficient carbon species such as carbocations and carbon radicals. Resonance stabilization of aromatic species' conjugated π systems is a major contributing factor in benzene's unusual stability. When electron donating groups (EDGs) with nonbonding electron pairs are located adjacent to carbocationic centers, stabilization occurs by resonance donation of an electron pair from the EDG to the carbocation. Likewise, π systems such as allylic and benzylic cases can delocalize positive charges by resonance.

Carbanions, conversely are stabilized by inductive withdrawal of electron density. Electron withdrawing groups (EWGs), even those without π systems, serve this purpose. However, increased stabilization results when an EWG withdraws electron density through resonance as well as induction. Finally, as we have shown, π systems such as allylic and benzylic cases can delocalize negative charges by resonance, thereby stabilizing the carbanion. Stabilization of radicals is also enhanced by allylic and benzylic π systems.

Solved Problems

Problem 4.1.
Given: Butane molecule.
Find: (1) Remaining Newman projections for the butane molecule, (2) calculate their energies.
Plan: (1) Draw the three Newman projections not presented in the text, (2) use Table 4.1 to calculate conformer energies.
Solve:

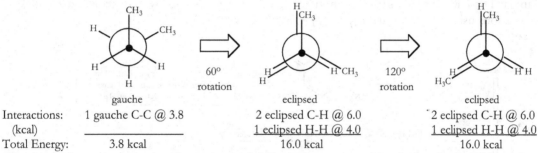

	gauche	eclipsed	eclipsed
Interactions: (kcal)	1 gauche C-C @ 3.8	2 eclipsed C-H @ 6.0 1 eclipsed H-H @ 4.0	2 eclipsed C-H @ 6.0 1 eclipsed H-H @ 4.0
Total Energy:	3.8 kcal	16.0 kcal	16.0 kcal

Check: (1) Newman projection drawn correctly, (2) conformer energies calculated using Table 4.1.

Problem 4.2.
Given: 2-Methylbutane molecule, sighting down the C2-C3 axis.
Find: Newman projections for the most and least stable conformations.
Plan: (1) Draw Newman projections for the most and least stable conformations, (2) use Table 4.1 to calculate conformer energies.
Solve:

a. most stable

gauche

b. least stable

eclipsed

Note: Eclipsed conformation drawn slightly offset to clearly show substituents.
Check: (1) Newman projection drawn correctly, (2) conformer energies calculated using Table 4.1.

Problem 4.3.

Given: 2-thiomethoxypropyl cation, $(CH_3)_2C^+(SCH_3)$

Find: (1) Resonance forms for $(CH_3)_2C^+(SCH_3)$ that show how delocalization of the positive charge takes place. (2) $(CH_3)_2C^+(SCH_3)$ stability relative to that of unstabilized 2° and 3° carbocations.

Plan: (1) Draw Resonance forms for $(CH_3)_2C^+(SCH_3)$, (2) compare this carbocation's stability to unstabilized 2° and 3° carbocations using resonance and hyperconjugation theory.

Solve:

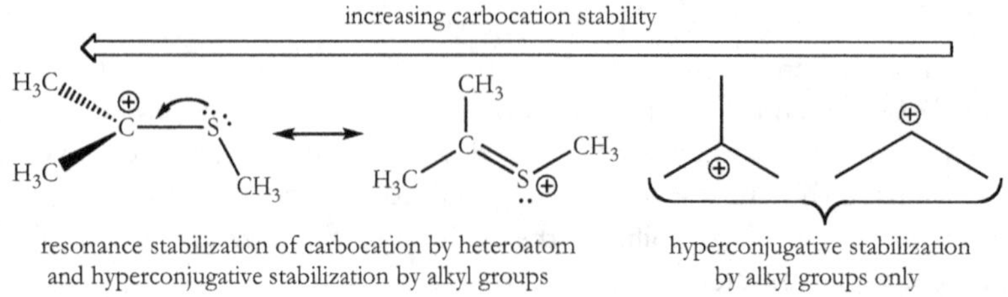

increasing carbocation stability

resonance stabilization of carbocation by heteroatom
and hyperconjugative stabilization by alkyl groups

hyperconjugative stabilization
by alkyl groups only

Check: (1) Carbocations drawn correctly, (2) resonance forms drawn that show charge delocalization, (3) carbocation stability trend consistent with theory.

Problem 4.4.

Given:

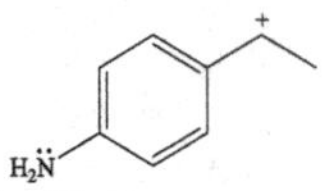

Find: (1) Resonance forms for benzylic carbocation depicted above. (2) Stability of this carbocation relative to that of unstabilized 2° and 3° carbocations.

Plan: (1) Draw Resonance forms, (2) compare these carbocation's stability to unstabilized 2° and 3° carbocations using resonance and hyperconjugation theory.

Solve:

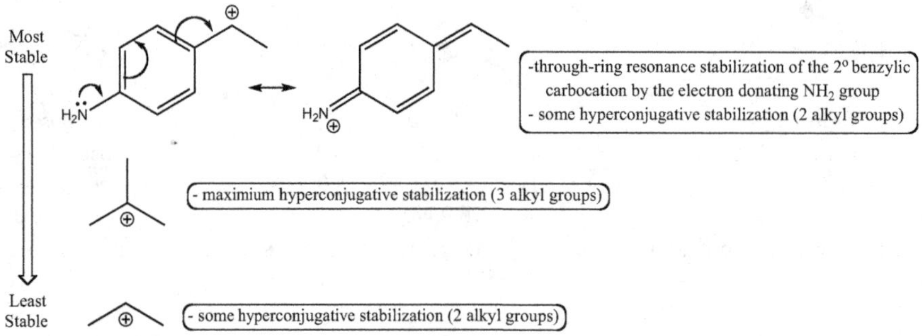

Most Stable

-through-ring resonance stabilization of the 2° benzylic carbocation by the electron donating NH_2 group
- some hyperconjugative stabilization (2 alkyl groups)

- maximimum hyperconjugative stabilization (3 alkyl groups)

Least Stable

- some hyperconjugative stabilization (2 alkyl groups)

Check: (1) Carbocations drawn correctly, (2) resonance forms drawn that show charge delocalization, (3) carbocation stability trend consistent with theory.

Problem 4.5.

Given:

Find: (1) Resonance forms for allylic and benzylic radicals depicted above. (2) Stability of these radicals relative to that of unstabilized 2° and 3° radicals.

Plan: (1) Draw Resonance forms, (2) compare these radical's stability to unstabilized 2° and 3° radicals using resonance and hyperconjugation theory.

Solve:

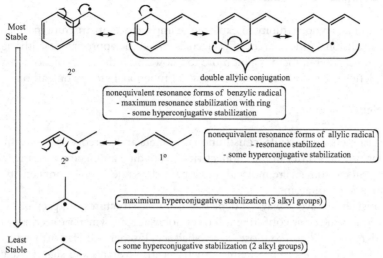

Most
Stable

2°

double allylic conjugation

nonequivalent resonance forms of benzylic radical
- maximum resonance stabilization with ring
- some hyperconjugative stabilization

2°   1°

nonequivalent resonance forms of allylic radical
- resonance stabilized
- some hyperconjugative stabilization

- maximium hyperconjugative stabilization (3 alkyl groups)

Least
Stable

- some hyperconjugative stabilization (2 alkyl groups)

Check: (1) Radicals drawn correctly, (2) resonance forms drawn that show electron delocalization, (3) radical stability trend consistent with theory.

Problem 4.6.
Given:

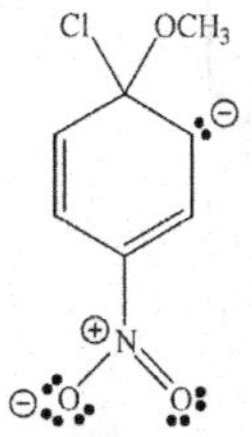

Cl   OCH$_3$

Find: (1) Resonance forms for Meisenheimer complex depicted above, (2) show stabilization of the carbanion by the π system of the aromatic ring and the EWG.
Plan: (1) Draw Resonance forms to show how the π system of the aromatic ring and the EWG stabilize the carbanion.
Solve:

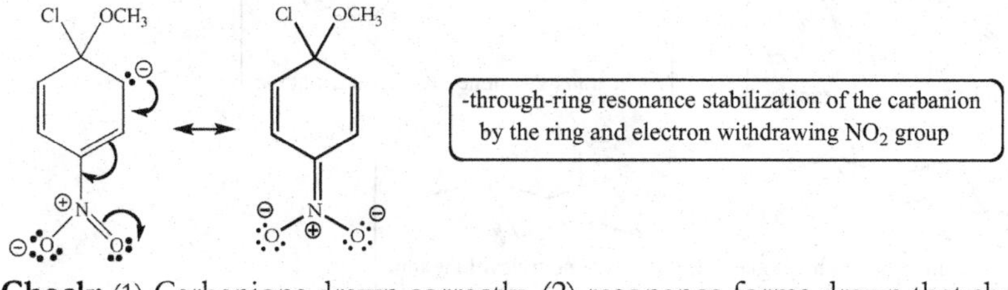

Cl   OCH$_3$ Cl   OCH$_3$

-through-ring resonance stabilization of the carbanion
by the ring and electron withdrawing NO$_2$ group

Check: (1) Carbanions drawn correctly, (2) resonance forms drawn that show charge delocalization occurs through the π system of the aromatic ring and the EWG.

Chapter 5 - Isomerism and Stereochemistry

Now that we've had an opportunity to explore the importance of structure in organic chemistry, the three-dimensional (3-D) nature of molecules should be appreciated. Although we introduced isomerism in Chapter 4, there is a bit more to know about the types of isomerism possible and how we define the spatial relationship between functional groups on carbon.

5.1 Types of Isomerism

Two types of isomers are possible, constitutional isomers and stereoisomers. Compounds which are constitutional isomers differ in the connectivity or bonding arrangement of the atoms which compose them. Stereoisomers are molecules that have the same atomic connectivity, but differ in the spatial orientation of atoms.

Organic chemists have several ways of depicting isomeric structures. In previous chapters, we explored a few cases like the chair conformation of cyclohexane, Newman projections and the use of dashes and wedges. In addition to those, as we shall see, Fischer and Haworth projections are used for molecules of special interest to biochemists such as carbohydrates and sugars like glucose. Figure 5.1 gives some examples.

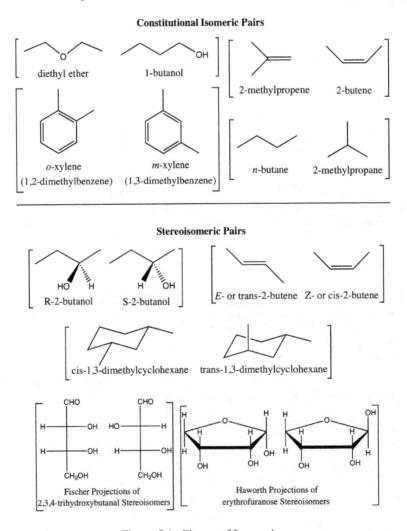

Figure 5.1. Types of Isomerism.

5.1a. Constitutional Isomerism

Compounds that differ in their atomic attachments but not in their overall molecular formulas are constitutional isomers. Typically, as in the examples shown in Figure 5.1, the compounds will have either different functional groups (ether: C-O-C linkage versus alcohol: C-O-H linkage) or substituents attached at different sites on the molecule (the methyl group in 2-methylpropane versus *n*-butane and 2-methylpropene versus 2-butene; the methyl groups in a 1,2 arrangement versus a 1,3 arrangement in xylene).

5.1b. Stereoisomerism

Molecules that do not have different constitutions and differ only in the spatial orientation of their atoms are considered stereoisomers. We draw stereoisomers to show clearly the three-dimensional relationship of the atoms at the molecular sites of interest. As we show in Figure 5.1, this stereoisomeric (3-D) orientation may be drawn using dashes and wedges (the 2-butanol isomers), depicted using lines to show the position of the substituents around a double bond (E/Z isomers of 2-butene) and by using axial and equatorial positions on the chair conformation of cyclohexane (cis and trans 1,3-dimethylcyclohexane isomers). Haworth projections (D-erythrofuranose) and Fischer projections (2,3,4-trihydroxybutanal) will be discussed later.

5.2 Handedness and Chiral Molecules

Hold your hands up so that the palms face each other. Your hands are mirror images. Now, try to overlay your hands and match your fingers up. You cannot overlay or superimpose them. This property is called handedness or chirality.

Chiral molecules are like hands. Figure 5.2 shows the mirror image pair of 2-chlorobutane stereoisomers.

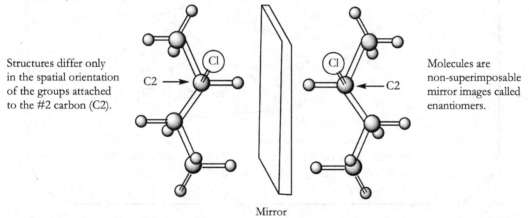

Structures differ only in the spatial orientation of the groups attached to the #2 carbon (C2).

Molecules are non-superimposable mirror images called enantiomers.

Mirror

Figure 5.2. 2-Chlorobutane Enantiomers.

Note that the number 2 carbon has four different groups attached to it: –H, -CH₃, -C₂H₅ and Cl. Such a carbon is termed a stereogenic center. The result is a molecule with no element of symmetry. The lack of symmetry is a requirement for a molecule to be chiral. To verify this, consider 2,2-dichlorobutane in Figure 5.3.

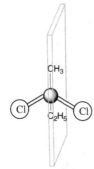

Figure 5.3. 2,2-Dichlorobutane.

The molecule possesses a plane of symmetry bisecting the carbon chain running between the chlorine atoms. Thus, 2,2-dichlorobutane is achiral. That is, the mirror images when overlaid will match everywhere. Let's get some practice with an example problem.

Example Problem 1. Predict whether 3-methylhexane is chiral.

Given: The molecule 3-methylhexane.
Find: Chirality center(s) present in 3-methylhexane, if any.
Plan: (1) Draw a 3-D representation of the molecule, (2) identify any carbons that bear four different groups, (3) verify that the molecule possesses no element of symmetry.
Solve:

1 chiral center, no symmetry element

Check: (1) Molecule is drawn correctly, (2) carbon 3 is the only carbon that bears four different groups, (3) the molecule possesses no element of symmetry.

Problem 5.1. Predict whether the following molecules are chiral.
  a. chlorofluoroethane b. 3-pentanol c. 2-chlorobutanal

5.3 Optical Activity

When you have a chiral molecule like one of the enantiomers of 2-chlorobutane (Figure 5.2), the possibility for another unique property arises, the ability to rotate a plane of polarized light. This property, known as optical activity, is one method for distinguishing between chiral stereoisomers.

Solutions of chiral molecules are placed in a device called a polarimeter. Light is passed through a polarizing slit, separating it into its electrical and magnetic components and allowing transmission of light in one orientation in the X-Y-Z coordinate plane through the solution. If the plane-polarized light is rotated, the solution exhibits optical activity. The degree of rotation observed at a specific temperature and wavelength, $[a]$, depends upon the solution concentration (c) in g/mL and the distance the light travels through the polarimeter (l), usually expressed in decimeters such that $[\alpha]^{25}{}_D = \alpha/l{\bullet}c$, where the subscript D refers to the wavelength associated with the D line of sodium measured at 25°C.

Rotation of the plane of polarized light occurs in either a clockwise (to the right or dextrorotatory (+)) or counterclockwise (to the left or levorotatory (-)) direction.

Finally, the (-) and (+) designations refer to stereoisomers with a particular configuration and have been used in the past in the nomenclature of enantiomeric molecules. For 2-chlorobutane, one enantiomer will rotate the plane of polarized light clockwise (+), while the other enantiomer (same concentration) will rotate the plane-polarized light an equal amount counterclockwise (-). Once identified by polarimetry, the enantiomers would be named according to its optical activity: (+) 2-chlorobutane and (-) 2-chlorobutane.

5.4 The Cahn-Ingold-Prelog (C-I-P) System – Establishing Priorities

The (+) and (-) polarimeter readings do not tell us to which chiral isomer the readings apply. Additional experimentation is required to identify which isomer gives the (+) or (-) reading. Once we know the 3-D structure of an isomer from X-ray and chemical correlations, we say we have the absolute configuration. Once the absolute configuration of a stereoisomer is established, if it gives a particular "+" reading in the polarimeter, its enantiomer will give an equal but "-" reading under the same conditions. Note, however, that we cannot deduce the absolute configuration or label the chiral molecule using the C-I-P system from a polarimeter reading unless the connectivity between the polarimeter reading and structure has been established by independent means.

Once absolute configuration is known, we may label the chiral molecule using the C-I-P system. To designate the absolute configuration of chiral molecules, we must first know how to prioritize functional groups attached to the stereogenic carbon. The system in use today, devised by the chemists Cahn, Ingold and Prelog, provides us with a means to do just that. By learning and applying the rules correctly, we can arrive at an unequivocal description of a chiral molecule. Let's apply the C-I-P rules to an enantiomer of 2-chlorobutane.

✓ **Rule 1:** Look at the atoms attached to the stereogenic center. Rank order the atoms with differing atomic masses. Precedence goes to atoms with the highest atomic mass.

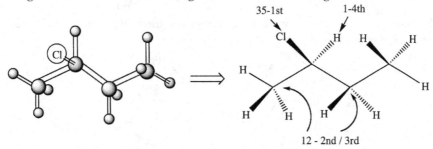

Chlorine is easily determined to be the highest priority; we assign it as 1st. Hydrogen has the lowest (4th) priority. The two carbon substituents may be assigned using Rule 2.

✓ **Rule 2:** If two substituents have the same atomic mass at the point of attachment, move to the next atom along the substituents' chains and apply rule 1. The methyl group, –CH₂-H, has only three hydrogens attached to the carbon, whereas the ethyl group, –CH₂-CH₃, has two hydrogens and a carbon atom attached. Based on Rule 1, then, the ethyl group is assigned 2nd in order of precedence and the methyl group is assigned 3rd.

✓ **Rule 3:** Multiple bond substituents are displayed as single bond units, but the atoms are doubled for substituents with double bonds and tripled for functional groups having triple bonds. Here are some examples:

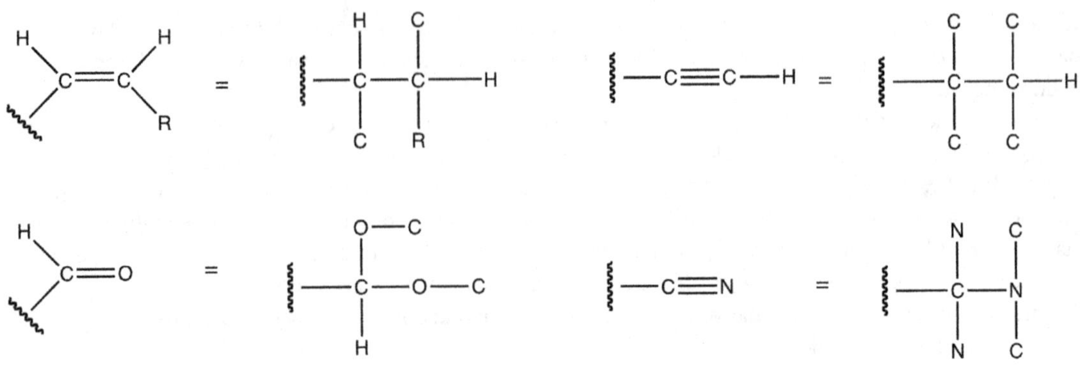

Problem 5.2. Rank the following substituents from lowest to highest precedence.
a. –SH   b. –Cl   c. -CN   d. –CHO

Problem 5.3. Rank the following substituents from lowest to highest precedence.
a. –COOH   b. –CH=CHCH₃   c. –C(CH₃)=CH₂ d. –CH₂OH

5.5 Configuration of Alkenes: cis/trans or E/Z?

The configuration of alkene geometric isomers such as those shown in Figure 5.1 may be assigned using the C-I-P rules. Remember that conformations are interconverted by bond rotation, but configurational changes involve bond breaking. To interconvert E and Z isomers, bonds must be broken and reassembled in a different geometry, since rotation about the double bond is restricted. Here is how the C-I-P rules are applied using the 2-butene isomers as examples:

❖ Draw the alkene structure so that you show how the substituents are arranged around the double bond:

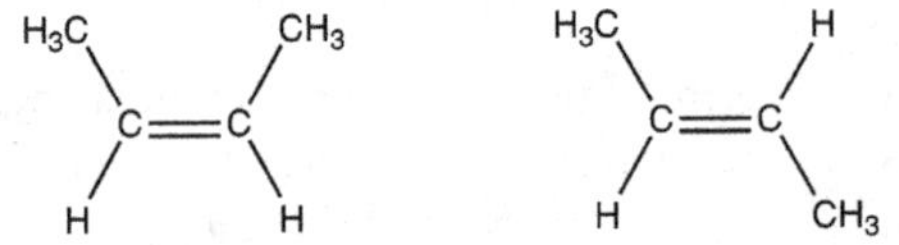

❖ Bisect the double bond vertically and prioritize the substituents as high or low priority using the C-I-P rules for each side of the bisecting line:

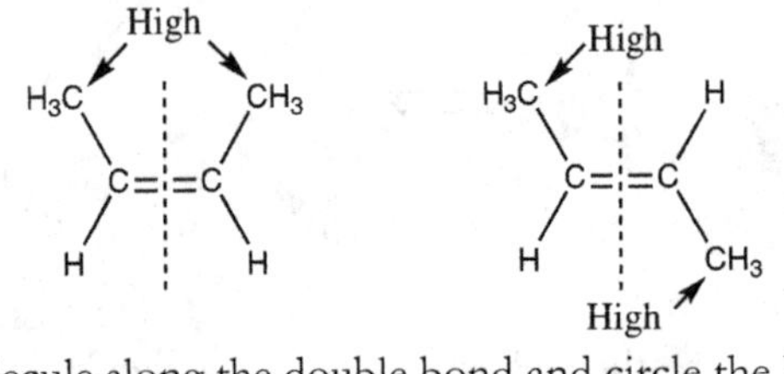

❖ Bisect the molecule along the double bond and circle the high priority groups:

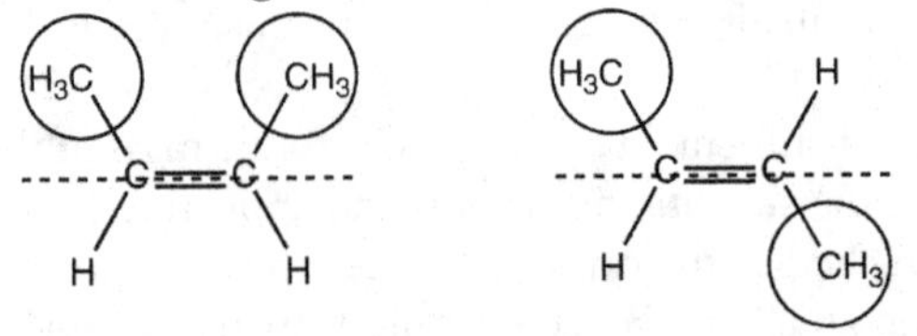

❖ If they are on the same side of the line drawn through the double bond, then we assign the alkene as a "Z" alkene; if the high priority groups reside on opposite sides of the line, then the alkene is said to be an "E" isomer.

❖ For disubstituted cases, "cis" **may** be used in place of "Z" and "trans" may be used in place of "E". Tri-and tetrasubstituted alkenes **must** be named using the E/Z system. Our alkenes, correctly named with the configuration prefix are:

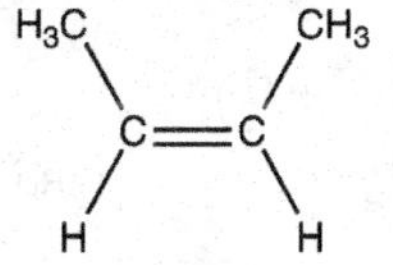

H₃C   CH₃   H₃C   H

$$C=C \qquad\qquad C=C$$

H   H   H   CH₃

Z-2-butene   E-2-butene

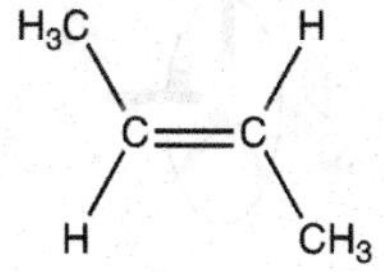

or   or

cis-2-butene   trans-2-butene

Problem 5.4. Predict whether the following alkenes are E or Z.

a.   b. 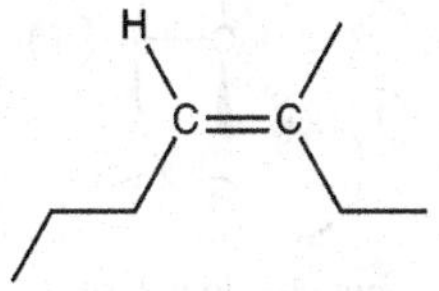   c.

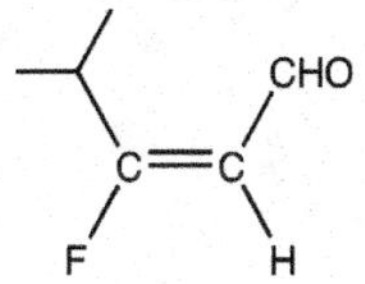

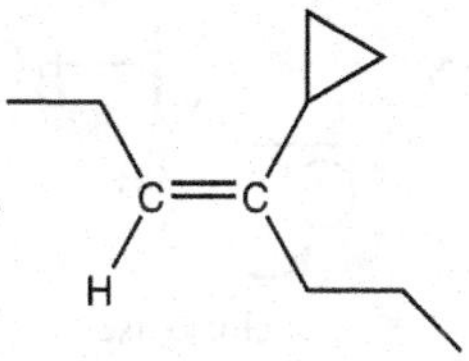

Naming molecules with multiple double bonds requires only that the E/Z prefix be used so as to account for each double bond unit separately by indicating both position and configuration in parentheses:

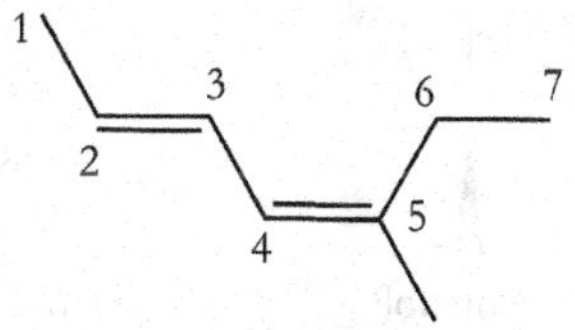

(2E, 4Z)-5-methylhepta-2,4-diene

5.6 Absolute Configuration of Stereoisomers Other Than Alkenes: the R/S convention

The absolute configuration of stereoisomers (other than alkenes) may be designated using the R (rectus or right) /S (sinister or left) convention. The R/S system describes a method for taking the substituents you prioritized from the C-I-P rules and permits you to determine the configuration around a chiral atom (usually carbon in organic chemistry). Here is how to do it using the 2-butanol isomers as examples:

❖ Draw the structure so that you show how the substituents are arranged around the chiral atom (making a model of the molecule will also help you to visualize the molecule):

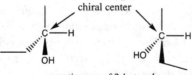

enantiomers of 2-butanol

❖ Prioritize the substituents from 1 - 4 using the C-I-P rules:

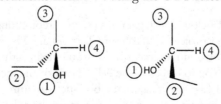

❖ Orient the molecule as if it were a steering wheel, placing the lowest priority group along the steering column axis facing away from you:

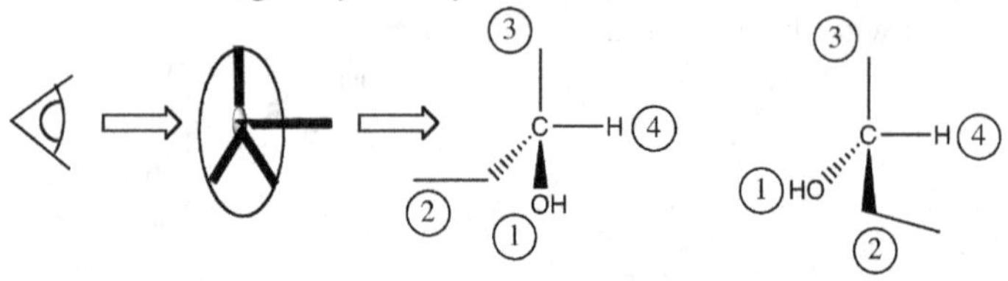

❖ Now trace a line from 1⇒2⇒3 noting the direction (either clockwise or counterclockwise) your path takes.

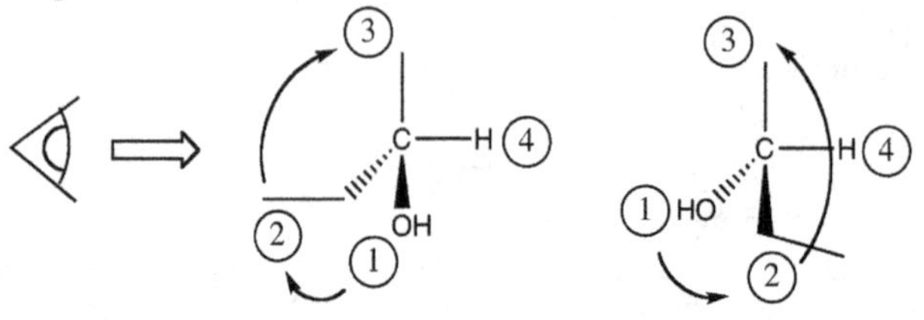

clockwise   counterclockwise

❖ If your path traced is clockwise, then the chiral center has an "R" configuration, whereas a counterclockwise path signifies an "S" configuration. Thus, the 2-butanol enantiomers have the following R/S designations:

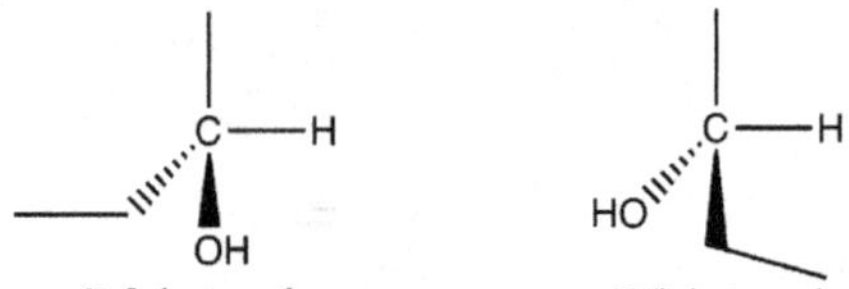

R-2-butanol   S-2-butanol

Problem 5.5. Assign R/S designations to the following enantiomers:

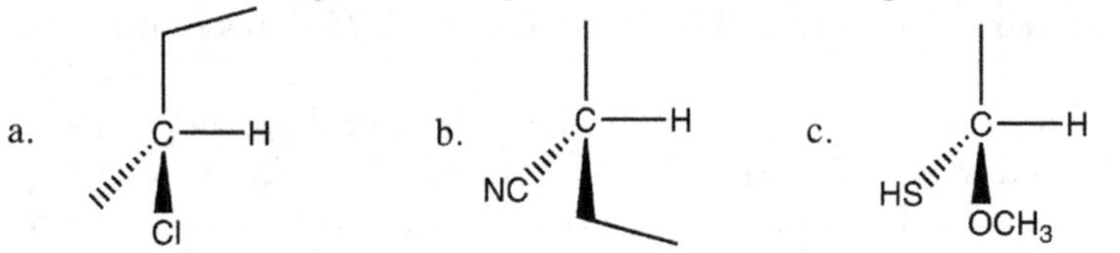

a.   b.   c.

5.7 Fischer Projections

These projections are two-dimensional representations of molecules that have a defined absolute configuration. Visualizing these molecules is challenging, but a good perspective of the 3-D structure will help you discern the absolute configuration from a 2-D structure. To illustrate, we'll use the Fischer projections of 2,3,4-trihydroxybutanal from Figure 5.1. A cursory glance at these molecules reveals 2 chiral centers. This will have implications for the types and numbers of stereoisomers possible and we will discuss that in due course. Here are some suggestions when working with Fischer projections:

✓ Structure 1 below is a staggered sawhorse representation of 2,3,4-trihydroxybutanal. The two-dimensional Fischer projection (structure 2) is obtained by rotating the sawhorse projection into an eclipsed conformation and flattening the substituents into the plane of the paper. (Again, making a model will enable you to visualize the 3-D nature of the molecule.) Here is how we render the Fischer projection in a 3-D manner called the bow-tie (structures 3 and 4). If we hold the center carbon-carbon bond of the bowtie fast, the substituents on the chiral carbons (bottom and top bow ties) may be rotated such that the lowest priority group (-H) faces away from you (recall the steering wheel model):

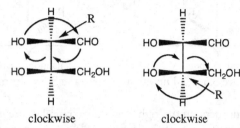

✓ Assign R/S to the stereocenters using the rules in Section 5.5.

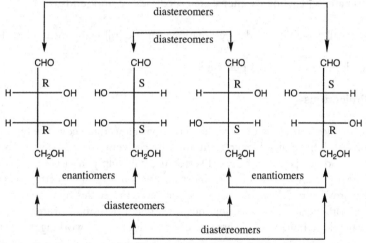

Notice that there are two stereocenters for our molecule. As a consequence, stereoisomers may be drawn that have different R/S designations, and as a result, we generate new stereoisomers called diastereomers (defined below). Let's look at all of the stereoisomeric possibilities:

Figure 5.4. Stereoisomeric Relationships in 2,3,4-trihydroxybutanal.

Note that the maximum number of stereoisomers can be determined by the formula 2^n (where n = # of stereocenters). In this case, $2^2 = 4$ possible stereoisomers.

The two enantiomeric pairs {(R,R), (S,S)} and {(R,S), (S,R)} as we indicated earlier are non-superimposable mirror images of one another. Notice that the enantiomers have the opposite stereochemical designation at all stereocenters. The stereoisomers outside of the enantiomeric pairings are diastereomers (stereoisomers which are not mirror images of one another). For stereoisomers to be diastereomers of one another, they must have at least one stereocenter with the same R/S designation and must have at least one stereocenter with the opposite designation.

Many years ago, biochemists adopted a separate system of nomenclature, the D/L system. Open chain sugars such as those in Figure 5.4 are identified as "D" sugars if the penultimate carbon (carbon 3 in our example) has the hydroxyl group on the right side of the Fischer projection. Therefore, in Figure 5.4, the (2R, 3R) and (2S, 3R) diastereomeric pair would be considered D sugars.

It follows that "L" sugars would have the same hydroxyl group on the left side of the Fischer projection. Note that this does not designate absolute configuration at all chiral centers, only that of the penultimate carbon.

5.8 Meso Compounds

As we mentioned in section 5.3, chiral molecules cannot possess an plane of symmetry. There are molecules which have multiple stereogenic centers, yet are achiral. Consider the stereoisomers of 1,2-dimethylcyclohexane in Figure 5.5:

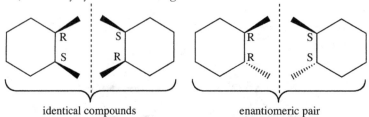

identical compounds   enantiomeric pair
Figure 5.5. 1,2-Dimethylcyclohexane Stereoisomers.

While the pair of trans enantiomers on the right are indeed chiral, there is something peculiar about the remaining cis pair. If we rotate one of the isomers 180°, we find the molecules to be identical! Put another way, if we draw a plane through the cis stereoisomer between the methyl groups, every point on the molecule reflects on itself; it is symmetric and therefore, achiral. Molecules with stereogenic centers and a plane of symmetry are called meso compounds.

Problem 5.6. Give all possible R and S designations for the following molecules. Identify the meso compounds.
a. 1,3-dimethylcyclopentane   b. 2,3-butanediol

5.9 Haworth Projections

These projections of cyclic organic molecules with a wedged edge are representations of molecules that have a defined absolute configuration by convention. Visualizing these molecules is challenging, but a good perspective of the 3-D structure will help you discern the absolute configuration from a Haworth structure. To illustrate, we'll use the Haworth projections of D-erythrofuranose from Figure 5.1. A cursory glance at these molecules reveals several chiral centers as well as interesting nomenclature. Assigning the absolute configuration of these molecules is exactly the same as we have done before. Here are some suggestions when working with Haworth projections:

➢ Redraw the ring to accurately reflect its 3-D structure. When you do this, the vertical bonds take on their axial or equatorial disposition.

> You may then assign priorities to functional groups and determine the R/S designations using the rules we learned in Sections 5.4 and 5.5.

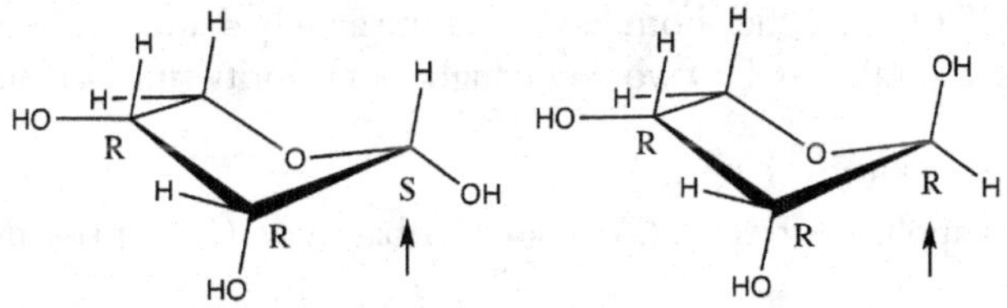

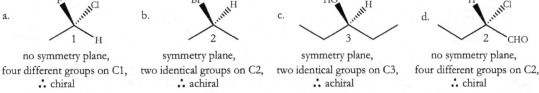

In addition to R/S designation, several other nomenclature items specific to sugars bear mentioning. On the structures above, the carbons marked by an arrow are identified by special greek symbols, α and β. When the hydroxyl group faces down in the Haworth Projection, the stereoisomer is designated as α, when facing up, β. A similar nomenclature system is used for steroid structures. The nomenclature prefix "*d*" refers to dextrorotatory; a molecule that rotates the plane of polarized light to the right (clockwise). Therefore, the names of the molecules above are α-*d*-erythrofuranose (structure on the left) and β-*d*-erythrofuranose (structure on the right). Levorotatory (*l*) is the designation assigned to molecules that rotate the plane of polarized light to the left (counterclockwise). It is of note, then, that *d*- and *l*- designations are not related to absolute configuration, but optical activity.

5.10 Summary

Isomerism is an important concept for the beginning organic chemistry student. Determining the absolute configuration of stereocenters within organic molecules by experimentation and the optical activity of a molecule are critical if we are to link three-dimensional structure to function and reactivity. This chapter has provided an introduction to understanding and identifying chirality, designation of R and S configurations using the Cahn-Ingold-Prelog system and stereochemical relationships between stereoisomers. Finally, this chapter has provided a glimpse of the various manners in which the stereoisomerism of organic molecules may be presented, e.g. E/Z designations of alkenes, cis/trans isomerism in cycloalkanes, as well as Fischer and Haworth projections.

Solved Problems

5.1.
Given: a. chlorofluoroethane b. 2-bromopropane c. 3-pentanol d. 2-chlorobutanal
Find: (1) Chirality center(s) present in a-d, (2) chiral molecules.
Plan: (1) Draw a 3-D representation of the molecules, (2) identify any carbons that bear four different groups, (3) verify that the molecules possess no plane of symmetry.
Solve:

a.	b.	c.	d.
no symmetry plane, four different groups on C1, ∴ chiral	symmetry plane, two identical groups on C2, ∴ achiral	symmetry plane, two identical groups on C3, ∴ achiral	no symmetry plane, four different groups on C2, ∴ chiral

Check: (1) Molecules are drawn correctly, (2) carbons identified that bear four different groups, (3) a and d possess no plane of symmetry, b and c do possess a symmetry plane.

5.2.
Given: a. –SH   b. –Cl   c. -CN   d. –CHO
Find: Rank a-d from lowest to highest C-I-P precedence.

Plan: Apply C-I-P rules to determine priority.
Solve: Rule 1: Assign priority based on atomic mass. Based on this, -Cl > -SH > -CN and CHO. Rule 2: For –CN and CHO, since they both have C as the attached atom, move to the next atom attached to the C. Based on this, -CHO would be higher in priority since O has a higher atomic mass than N.
Rank order: -Cl> -SH > -CHO > -CN.
Check: (1) C-I-P rules applied correctly, (2) rank order based on C-I-P priorities.

5.3. Rank the following substituents from lowest to highest precedence.
Given: a. –COOH   b. –CH=CHCH$_3$ c. –C(CH$_3$)=CH$_2$ d. –CH$_2$OH
Find: Rank a-d from lowest to highest C-I-P precedence.
Plan: Apply C-I-P rules to determine priority.
Solve: Rule 1 does not apply, since all substituents contain the C atom as the point of attachment. Rule 2: Move to the next atom attached to the C. Based on this, a. and d. have an O attached to the C, whereas b. and c. have a C attached. Therefore a. and d. are higher in priority.
Rule 3: Multiply bonded substituents must be expanded so they will have higher priority. Here are the expanded sets:

Increasing priority

3 oxygens attached   1 oxygen attached   3 carbons attached   2 carbons attached

Rank order: -COOH > –CH$_2$OH >–C(CH$_3$)=CH$_2$ > –CH=CHCH$_3$
Check: (1) C-I-P rules applied correctly, (2) rank order based on C-I-P priorities.

5.4.
Given:

a.   b.   c.

Find: Predict whether the following alkenes are E or Z.
Plan: (1) Apply C-I-P rules to determine priority, (2) identify alkenes as E or Z.
Solve: (1) Bisect the alkene with a vertical line, use C-I-P rules to determine priority of the substituents on each side of the bisected alkene. (2) Label the high priority substituents w/an H, the low priority substituents w/an L. (3) Bisect the alkene with a horizontal line. If the H substituents are on the same side of the horizontal line, the alkene is "Z". If not, then the alkene is "E".

a.   b.   c.

E   Z   Z

Check: (1) C-I-P rules applied correctly, (2) alkenes identified as either E or Z.

5.5.

Given:

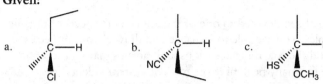

a.   b.   c.

Find: Assign R/S designations to any chiral centers in a-c.

Plan: (1) Identify chiral centers, (2) apply C-I-P rules to determine priority of substituents, (2) determine R/S designations.

Solve: (1) Use C-I-P rules to determine priority of the substituents from 1(highest)-4 (lowest), (2) reorient the molecules so that the lowest priority group is facing away from you (behind the plane of the paper), (3) trace an arc from the substituent labeled 1 → 2 → 3. If the direction of travel is clockwise, the chiral center is "R".

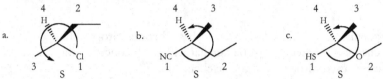

a.   b.   c.

Check: (1) C-I-P rules applied correctly, (2) molecules identified as "S".

5.6.

Given: a. 1,3-dimethylcyclopentane   b. 2,3-butanediol

Find: Give all possible *R* and *S* designations for a and b. Identify the meso compounds.

Plan: (1) Draw structures, (2) identify chiral centers and elements of symmetry, (3) apply C-I-P rules to determine priority of substituents, (4) determine R/S designations and meso compounds.

Solve: (1) Draw 3-D structure for a; Fischer projection for b, orienting the molecules so that the lowest priority group is facing away from you. (2) Identify chiral centers and any elements of symmetry. (3) Use C-I-P rules to determine priority of the substituents on chiral centers, draw in planes of symmetry. (4) Designate R/S for chiral centers, meso compounds for those with a plane of symmetry:

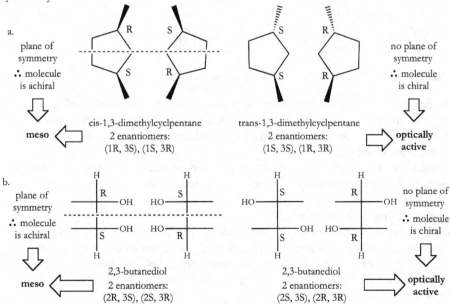

Check: (1) Molecules drawn in 3-D correctly, (2) chiral centers/planes of symmetry identified, (3) C-I-P rules applied correctly for the chiral centers, (4) R/S designated, meso compounds identified.

Chapter 6 - Function of Reagents in Organic Chemistry

The preceding chapters have largely been devoted to one of the important aspects of organic chemistry – structure. As we have seen, unless you are able to visualize and draw chemical species in three dimensions, it is difficult to understand the role molecules, ions and radicals play in a reaction.

We now turn to another challenge for many beginning organic chemistry students – the function of typical reactants and reagents used in organic reaction processes. How can you tell, for instance, how an alkene is likely to react with another species? What role does an acid or base play in an organic reaction? Why do nucleophiles and electrophiles interact? Herein, we will point out some common definitions and concepts that, in concert with structure, will improve your ability to identify the role a reagent may play in a given organic chemical process. Then in Chapter 7, we will apply this newly found understanding to explore how organic reactions occur.

6.1 Acids and Electrophiles

Beginning chemistry students are introduced to two acid-base theories, proposed by Arrhenius and Lewis. Organic chemistry students must be familiar with both concepts since many different types of acids and bases are used throughout organic chemistry.

➢ Acids
Arrhenius defined an acid as a species which donates an H^+ ion in solution:

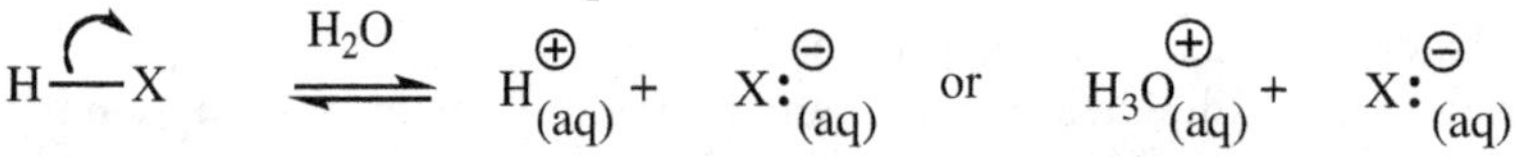

$$H \overset{\frown}{\text{---}} X \quad \overset{H_2O}{\rightleftharpoons} \quad H^{\oplus}_{(aq)} + X:^{\ominus}_{(aq)} \quad \text{or} \quad H_3O^{\oplus}_{(aq)} + X:^{\ominus}_{(aq)}$$

We show the dissociation of the acid in an aqueous medium by drawing a curved arrow to indicate movement of the electron pair in the H-X bond toward X, leaving an H^+ ion and X:⁻. Many Arrhenius (also known as Brønsted-Lowry) acids fit this definition; some examples are shown in Table 6.1.

Lewis expanded the definition of an acid to all species which can accept electron pairs. Therefore, any molecule or ion that is deficient in electron density can function as a Lewis acid. For example:

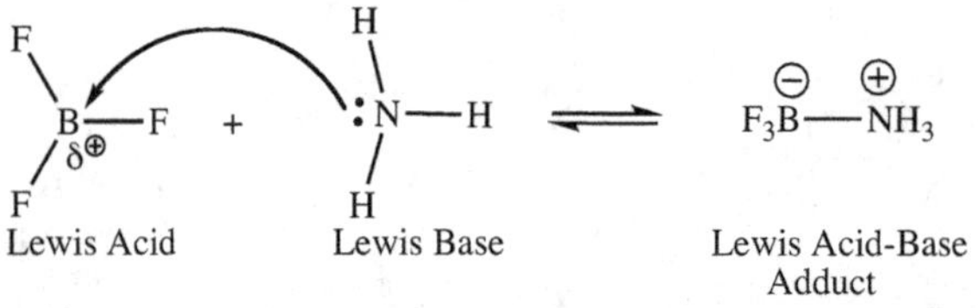

Here, the three fluorine atoms attached to the already electron-deficient boron atom create a substantial partial positive charge at boron. Thus, the electron-deficient boron trifluoride functions as a Lewis acid and receives an electron pair (depicted by a curved arrow) from ammonia, the Lewis base, to form a Lewis acid-base adduct.

If we now look back at the dissociation of HX, the H^+ ion qualifies as a Lewis acid based on our definition. **This is true of all Arrhenius acids; they are Lewis acids.**

➢ Electrophiles – "Electron-loving" species
As we noted in the reaction above between the Lewis acid BF_3 and Lewis base NH_3, the electron deficiency of boron is what causes it to be an electron pair acceptor or electron-loving. Therefore, molecules with a partial positive charge arising from a bond dipole or ions with full positive charges can be electrophiles – electron pair acceptors. Table 6.1 lists some examples.

Table 6.1. Acids and Electrophiles

Arrhenius Acids	Lewis Acids	Electrophiles
H^+, H_3O^+ sources: H-X (X=F, Cl, Br, I) $^+NH_4$, HNO_3, H_2SO_4, H_3PO_4	H^+, H_3O^+ sources: H-X (X=F, Cl, Br, I) $^+NH_4$, HNO_3, H_2SO_4, H_3PO_4	H^+, H_3O^+ sources: H-X (X=F, Cl, Br, I) $^+NH_4$, HNO_3, H_2SO_4, H_3PO_4
	BF_3, $AlCl_3$	BF_3, $AlCl_3$
	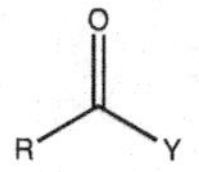 (Y= H, OR, R, Cl, Br)	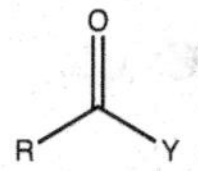 (Y= H, OR, R, Cl, Br) (Carbon of C=O is electrophilic)
	R-X (X=electronegative element) R^+	R-X (X=electronegative element) R^+

Problem 6.1 Identify the partial positive charge locations for the following Lewis acids/electrophiles:
a. BF_3 b. $AlCl_3$ a. H_2CO

6.2 Bases and Nucleophiles

➤ Bases

Arrhenius defined a base as a species which donates hydroxide ion, OH^-, in solution:

We show the dissociation of the base in an aqueous medium by drawing a curved arrow to indicate movement of the electron pair in the M-OH bond toward OH, leaving an M^+ ion and $HO:^-$. Many Arrhenius (also known as Brønsted-Lowry) bases fit this definition; some examples are shown in Table 6.2.

Lewis expanded the definition of a base to all species which can donate electron pairs. Therefore, any molecule or ion that has extra electron density, eg. contains an electron pair can function as a Lewis base. Turning again to our previous example:

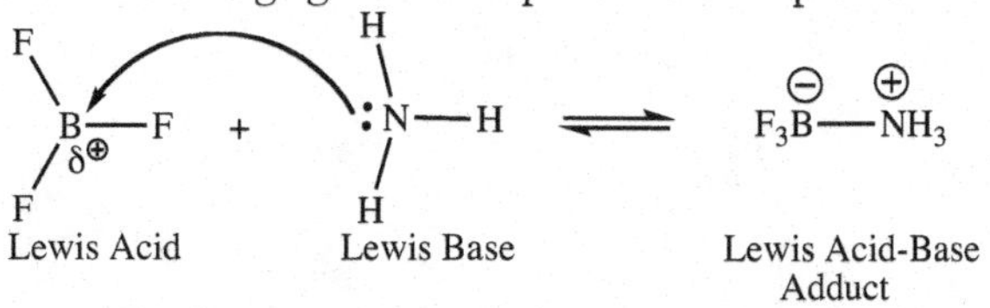

The non-bonding electron pair of the nitrogen atom is attracted to the partial positive charge at boron of the electron-deficient boron trifluoride. Ammonia, the Lewis base, donates its non-bonding electron pair (indicated by a curved arrow) to form a Lewis acid-base adduct.

If we now look back at the dissociation of MOH, the $HO:^-$ ion qualifies as a Lewis base. **This is true of all Arrhenius bases; they are Lewis bases.**

➤ Nucleophiles – "Nucleus-loving" species.

As we noted in the reaction above between the Lewis acid BF_3 and Lewis base NH_3, the non-bonding electron pair of the nitrogen atom is what causes it to be an electron pair donor or nucleus-loving. Therefore, molecules with an electron pair (either bonding or non-bonding) or ions with a full negative charge can be nucleophiles – electron pair donors. Table 6.2 lists some examples.

Table 6.2. Bases and Nucleophiles

Arrhenius Bases	Lewis Bases	Nucleophiles
HO:⁻ sources:	HO:⁻ sources:	HO:⁻ sources:
M-OH (M=H, Li, Na, K); Ca(OH)$_2$	M-OH (M=H, Li, Na, K); Ca(OH)$_2$	M-OH (M=H, Li, Na, K); Ca(OH)$_2$
R$_2$N:⁻, :NR$_3$ (R=H, alkyl, aryl)	R$_2$N:⁻, :NR$_3$ (R=H, alkyl, aryl)	R$_2$N:⁻, :NR$_3$ (R=H, alkyl, aryl)
X:⁻ (X=F, Cl, Br, I)	X:⁻ (X=F, Cl, Br, I)	X:⁻ (X=F, Cl, Br, I)
RS:⁻, RO:⁻	RS:⁻, RO:⁻	RS:⁻, RO:⁻
R:⁻	R:⁻	R:⁻
	R$_2$C=CR'$_2$	R$_2$C=CR'$_2$
	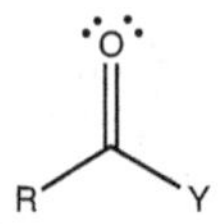 R—C(=Ö:)—Y (Y= H, OR, R, Cl, Br) (nonbonding e⁻ pair on C=O)	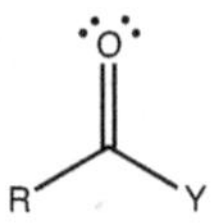 R—C(=Ö:)—Y (Y= H, OR, R, Cl, Br) (nonbonding e⁻ pair on C=O)

Problem 6.2 Identify nonbonding electron pairs and/or charge locations for the following Lewis bases/nucleophiles:

a. HO⁻   b. CH$_3$SH   a. H$_2$CO

6.3 Summary

Understanding the function of acids, bases, electrophiles and nucleophiles is essential to prediction of reaction products and how reactions take place. The Arrhenius definition defines acids as proton donors and bases as proton acceptors; we now have a better appreciation for that somewhat limited definition. The broader Lewis definition in which electron poor species (acceptors) are Lewis acids or electrophiles, while electron-rich species are Lewis bases or nucleophiles, encompasses all Arrhenius acids and bases. The concept that reactions occur between nucleophiles and electrophiles is fundamental and figures prominently in organic chemistry. In the next chapter, we will use reagent function as a basis for understanding the step-by-step processes by which molecules react with acids, bases, electrophilic and nucleophilic species in some important organic reaction mechanisms.

Solved Problems

6.1
Given: a. BF_3 b. $AlCl_3$ a. H_2CO
Find: Identify the partial positive charge locations for the following Lewis acids/electrophiles a-c.
Plan: (1) Draw reagents, (2) identify electrophilic/positive charge sites.
Solve:

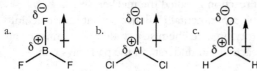

For a-c, electrophilic sites are indicated by δ^+. For BF_3 and $AlCl_3$, each B-F and Al-Cl bond contributes to increasing the δ^+ on the central atom.
Check: (1) Reagents drawn correctly, (2) identified electropositive sites on molecules.

6.2
Given: a. HO^- b. CH_3SH a. H_2CO
Find: Identify nonbonding electron pairs and/or charge locations for the following Lewis bases/nucleophiles in a-c.
Plan: (1) Draw reagents, (2) identify nucleophilic/negative charge sites.
Solve:

a. H—O:⊖   nucleophilic site (e⁻ pairs)
b. H₃C—S: H
c. nucleophilic site (e⁻ pairs)

Check: (1) Reagents drawn correctly, (2) identified nucleophilic/negative charge sites on molecules.

Chapter 7 - Step by Step: How Reactions Occur

The preceding six chapters have been devoted to fundamental principles, which allow us to draw chemical structures correctly, understand stability and reactivity trends in organic molecules as well as the functions of various reagent types. Now it is time to apply structure and function to get at how molecules and ions react. There are also radical processes, but they are beyond our scope here. The step-by-step process we describe for a given reaction is called the mechanism.

Mechanisms are educated guesses (supported by experimental evidence) about how reactant species interact to form new products. As an introductory organic chemistry student, you must bring to bear all your knowledge of structure, function and reactivity to understand the pathways that reactants take during their transformation to products.

Our goal in this chapter is to highlight some of the more illustrative and important reaction pathways, as an exhaustive review of every mechanism introductory organic chemistry students may encounter is simply not possible. To do this, we will dissect these pathways by our systematic problem-solving approach so that you will be able to:

✓ Correctly draw the structures of the reactive species that participate in the reactions,
✓ Identify the function of all species involved, and
✓ Correctly show electron flow between reactants using the "curved arrow formalism".

Following these steps will permit you to arrive at plausible reaction mechanisms for the transformations while giving you insight into how reaction trends may be applied to cases you have not been exposed to.

The mechanisms we will examine are:

- o Acid-Base Reactions (proton transfers)
- o Alkene Additions
- o Nucleophilic Substitutions
- o Eliminations
- o Electrophilic Aromatic Substitutions
- o Additions/Substitutions to Carbonyl Species
- o Carbonyl α-Substitutions
- o Conjugate Additions to α, β-Unsaturated Carbonyls
- o Oxidation of Alcohols

7.1 Acid-Base Reactions / Proton Transfers

Because many reactions in organic chemistry require acids or bases to catalyze or promote them, proton transfers are important to understand. As we learned in Chapter 6, acids are defined as electron deficient species that accept an electron pair from a base. When this happens, the electron pair bonds with a proton, transferring it to the base (donor species). Proton transfers can occur between molecules or ions as the example below shows.

Example Problem 1. Write the mechanism for the following proton transfer:

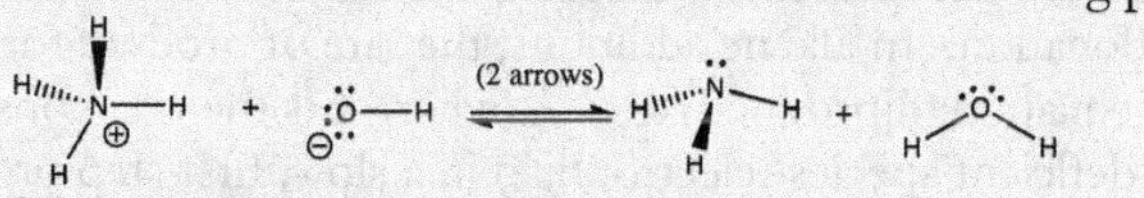

Given: Acid-Base reaction as shown above, 2 arrows needed.
Find: Step-by-step process depicting the transfer of a proton.
Plan: (1) Identify the acid (proton donor or e⁻ pair acceptor), the base (proton acceptor or e⁻ pair donor) and leaving group.
(2) Use the curved arrow formalism to indicate donation of the e⁻ pair from the base to the acid.
(3) Use the curved arrow formalism to indicate donation of the e⁻ pair from the N-H bond to the nitrogen atom.
Solve:

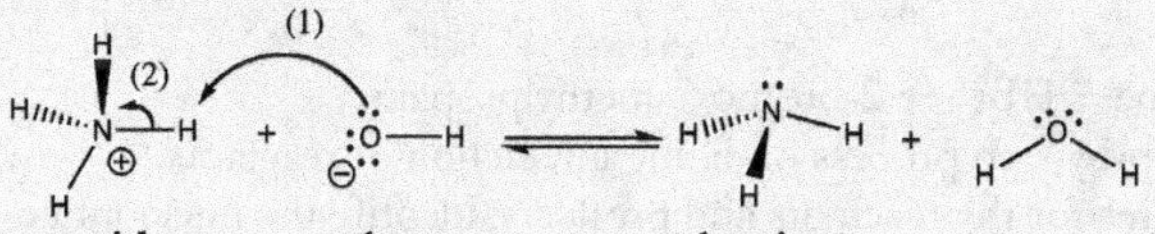

The arrows drawn in this manner show:
(1) The hydroxide oxygen forms a new H-O bond leading to the production of H_2O,
(2) Receipt of the N-H bonding electron pair by nitrogen leads to the production of NH_3.
Check: (1) Acid/Base properly identified, (2) Curved arrow formalism used correctly such that the observed products result from the unidirectional movement of electron pairs in bond formation and bond breaking.

Now, we can use this knowledge to predict the outcome of similar processes.

Problem 7.1. Analyze the reactions below with electron pushing arrows and predict the outcome.

(a)

leaving group

(b)

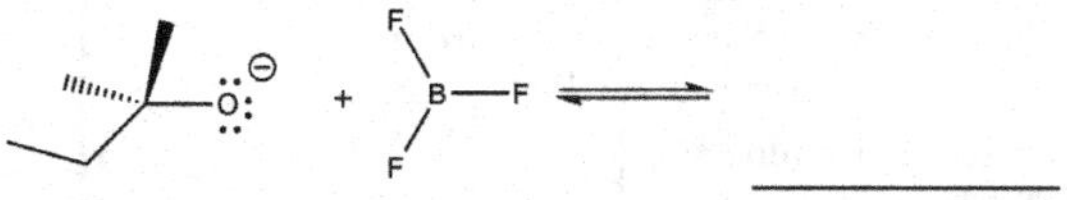

Lewis acid-base adduct

(c)

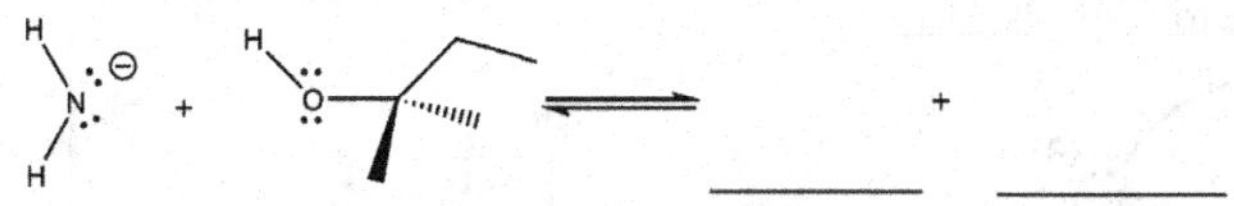

leaving group

Problem 7.2. Analyze reaction (a) with electron pushing arrows. Then, complete the mechanism for (b) and predict the outcome.

(a)

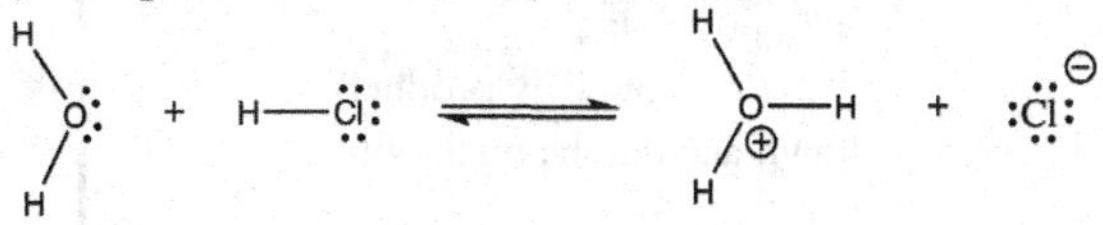

(b)

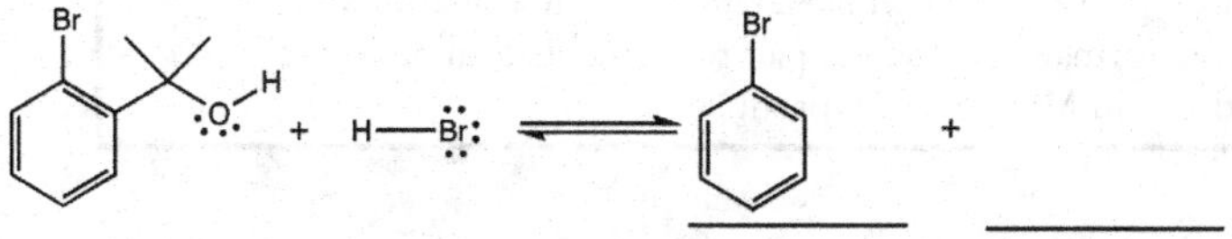

finish structure   leaving group

69

7.2 Alkene Additions

Double bonds are nucleophilic; that is, the electron pair contained in the π bond is available under the right conditions for donation. In alkene additions, the rate of product formation (d[product]/dt) is given by the equation: d[product]/dt = k (where k is the rate constant)[alkene]. The alkene attacks an electron-deficient species (electrophile) in a slow, first step producing a carbocation intermediate (Chapter 4) and a leaving group. The carbocation intermediate is then captured in a subsequent fast step by the leaving group to yield an addition product. How does this occur? Let's look at an example problem.

> Addition of H-X (X-Cl, Br, I) to an alkene

Example problem 2. Show the mechanism for the addition reaction between 2-methylpropene and H-Br to give 2-bromo-2-methylpropane.

Given: 2-methylpropene + HBr → 2-bromo-2-methylpropane.
Find: Mechanism (step-by-step process outlining interaction of reactants.
Plan: (1) Draw structures for the reactants and products; identify the nucleophile and electrophile.
(2) Write the stepwise mechanism showing all reactants, intermediates and products, using curved arrow formalism to show e⁻ pair movements.
Solve:
(1) Overall reaction:

nucleophile   electrophile

(2) Mechanism (2 steps, 3 arrows needed):

step 1: <u>carbocation formation</u>

alkene nucleophile donates e⁻ pair   most stable   leaving
to HBr electrophile   carbocation   group
                                             forms (Chp 4)

{ This step is normally the rate-determining step
(slowest step) in alkene additions of this type }

step 2: <u>carbocation capture to form alkyl bromide</u>

Br:- anion donates e⁻ pair
to carbocation   most highly substituted
                                  alkyl halide forms
                                  (Markovnikov product)

Check: (1) Structures for all reactants and products drawn accurately; nucleophile and electrophile identified.
(2) Stepwise mechanism proposed showing all reactants, intermediate and product. Curved arrow formalism shows e⁻ pair movements leading to the most stable carbocation and Markovnikov product.

Problem 7.3. Analyze the following reactions with electron pushing arrows and predict the outcome.

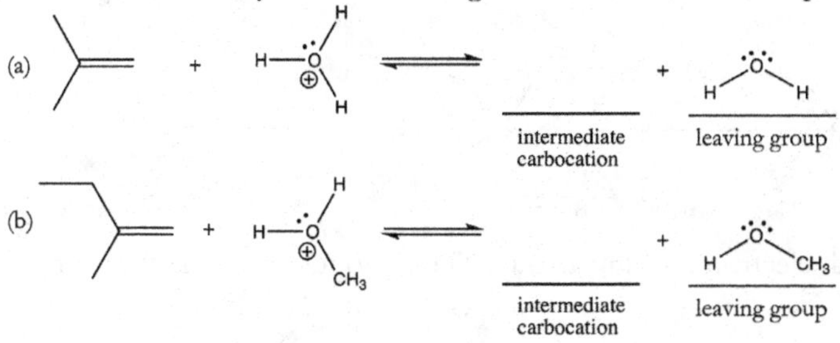

(a)

```
           H
           |
  \       H—O:        ⇌                    +      :O:
   >=              ——————  +  ——————            H      H
  /    +      ⊕
              |               intermediate      leaving group
              H               carbocation
```

(b)

```
           H
           |
  \       H—O:        ⇌                    +      :O:
   >=              ——————  +  ——————            H      CH₃
  /    +      ⊕
              |               intermediate      leaving group
             CH₃              carbocation
```

The product stereochemistry observed for alkene additions, which proceed through carbocation intermediates, is governed by the trigonal planar geometry of the carbocation (Chapter 4). As Figure 7.1 shows, carbocation capture by a nucleophile may occur by attack on either face of the empty p orbital. Even in cases where the resulting product molecules are chiral, the lack of nucleophilic preference toward either carbocation face leads to a mixture of enantiomeric products, usually in equal amounts. Such mixtures are termed racemic and are not optically active.

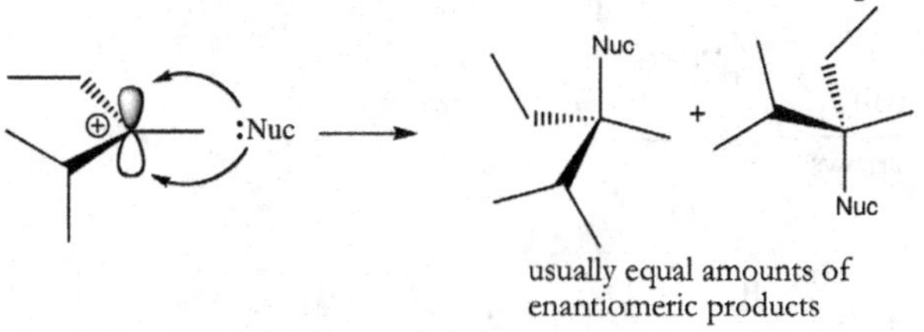

usually equal amounts of
enantiomeric products

Figure 7.1. Alkene Addition Product Mixture.

➤ Carbocation Rearrangements

The cases we have examined thus far featured formation of the most stable carbocation. What happens if the carbocation that forms during an alkene addition reaction is not the most stable possible? For example, consider the reaction below:

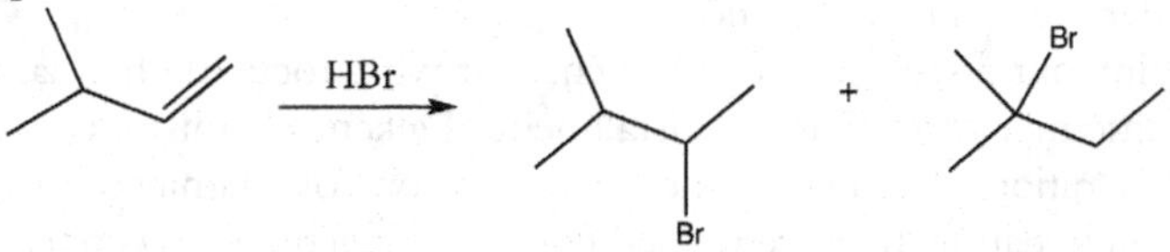

(A) 50%   (B) 50%

Analysis of the products reveals the alkyl halide A results from normal Markovnikov addition of HBr to the alkene. The remaining product, B, forms from a rearrangement. Here is how product B arises.
Step 1:

2° carbocation

From Chapter 4, recall that a 2° carbocation is less stable than a 3° carbocation. When a hydride or other group on a carbon (#3 carbon) adjacent to the carbocation center (#2 carbon) is available to migrate with its electrons, a new 3° carbocation forms at carbon #3.
Step 2:

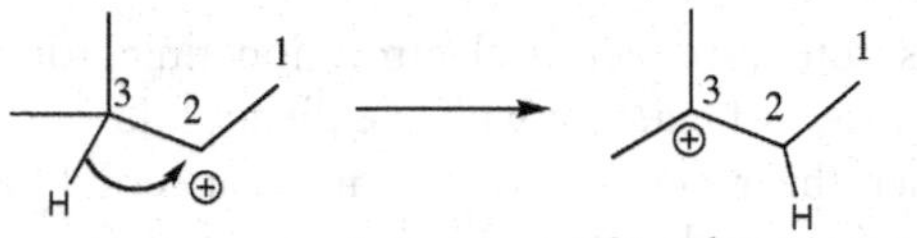

3° carbocation

Now, the bromide anion captures the more stable carbocation, producing the 2-bromo-2-methylbutane in step 3:

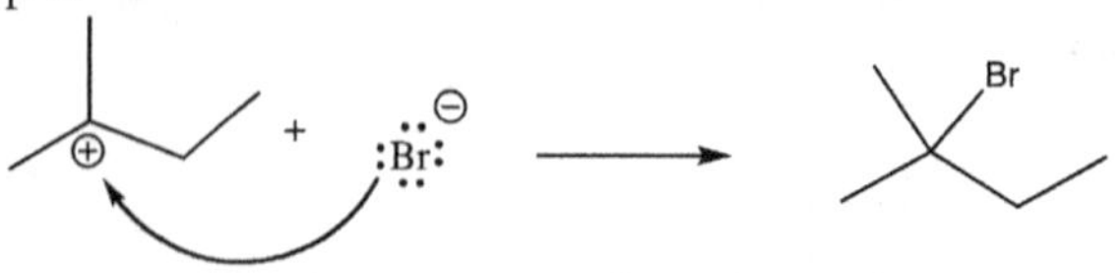

Problem 7.4. Analyze reaction (a) with electron pushing arrows. Then, propose a mechanism for (b).

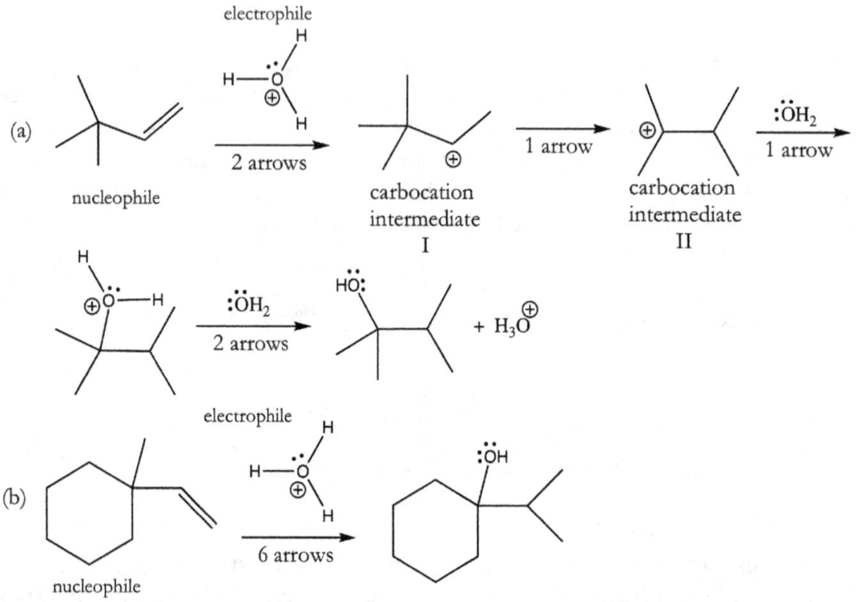

> ➢ Diels-Alder cycloaddition reactions

Up to this point, our discussion of alkene chemistry has focused on polar processes that generate carbocation intermediates. Our last installment of alkene chemistry concerns a unique class of reaction – the cycloaddition. The Diels-Alder reaction is one of a family of processes that is neither polar nor radical in nature; they are termed concerted reactions. By concerted we mean that electron flow between the reactants occurs simultaneously without the generation of intermediates of any type. The general reaction involving a dienophile (an alkene usually activated with an electron withdrawing group or EWG) and a 1,3-diene to give a mixture of enantiomeric cyclohexene products is shown below:

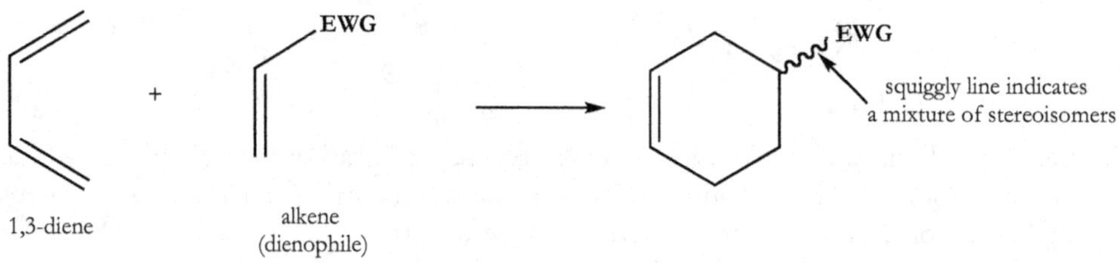

The 1,3-diene may also be substituted; frequently electron donating groups (EDGs) are attached which further activate the diene toward reaction with the dienophile.

As we mentioned earlier, this reaction proceeds in a concerted manner. Let's examine how such a process takes place by looking at the mechanism for the reaction between 1,3-butadiene and propenal:

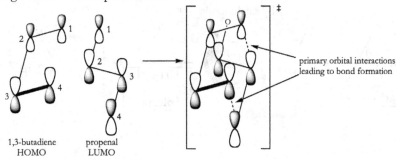

Notice that the electrons involved in this process are found in the π system. The 1,3-diene contributes 4π e⁻ and the dienophile 2π e⁻.

Molecular orbital theory provides a rationale for understanding the mechanism of the Diels-Alder cycloaddition. The molecular orbital interactions for the reaction between 1,3-butadiene and propenal through route A are depicted below:

1,3-butadiene HOMO   propenal LUMO   primary orbital interactions leading to bond formation

As the diagram shows, the propenal LUMO orients itself as the lobes on carbons 3 and 4 move underneath the lobes on carbons 1 and 4 of the 1,3-butadiene HOMO to allow for maximum overlap of similar shaded orbital lobes. These primary orbital interactions lead to bond formation required to produce the six-membered ring.

Problem 7.5. Predict the products for reactions in (a) and (b).

(a)

(b)

7.3 Nucleophilic Substitutions: S$_N$2 and S$_N$1 Reactions

Substitution reactions generally follow the form shown below:

$$\text{Nuc:}^- \quad + \quad \text{R-X} \quad \rightarrow \quad \text{Nuc-R} \quad + \quad \text{:X}^-$$
$$\text{or} \quad (\text{Nuc:}) \qquad\qquad (\text{Nuc}^+\text{-R})$$

where one reactant, the nucleophile (Nuc:$^-$) displaces part of the other reactant (X is the leaving group displaced from R-X in this example). While polar and radical reactions of this type may occur, we will limit our discussion in this section to two specific polar processes: substitution, nucleophilic, bimolecular [2nd order], (S$_N$2) and substitution, nucleophilic, unimolecular [1st order], (S$_N$1). It is important to stress here that the S$_N$2 and S$_N$1 processes we will discuss are the limiting cases along a continuum of nucleophilic substitutions.

7.3a. The Bimolecular S$_N$2 Process

Experimental kinetics studies indicate that S$_N$2 process are characterized by second order kinetics given by Rate = k[Nuc][R-X]. The rate law is consistent with two species in the transition state so the process is bimolecular. The effects that the nucleophile (Nuc:) and substrate (R-X) have on the reaction rate are substantial and will be discussed in detail shortly. Likewise, leaving group abilities and solvent contributions to the rate of S$_N$2 reactions are important and must be considered to get a complete picture of the factors that influence this important process in organic chemistry. Let's examine the reaction between potassium iodide and bromomethane below:

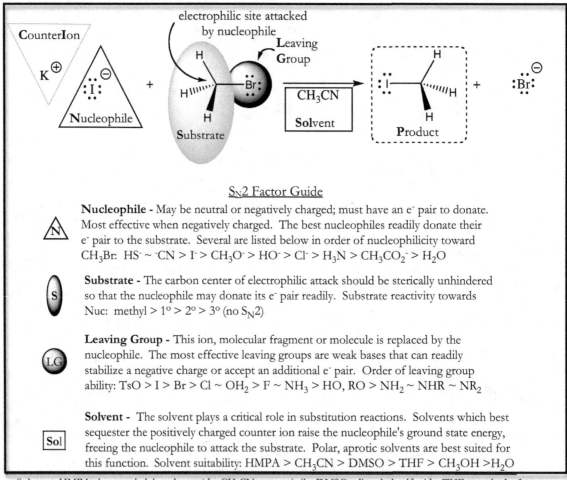

S$_N$2 Factor Guide

Nucleophile - May be neutral or negatively charged; must have an e$^-$ pair to donate. Most effective when negatively charged. The best nucleophiles readily donate their e$^-$ pair to the substrate. Several are listed below in order of nucleophilicity toward CH$_3$Br: HS$^-$ ~ $^-$CN > I$^-$ > CH$_3$O$^-$ > HO$^-$ > Cl$^-$ > H$_3$N > CH$_3$CO$_2^-$ > H$_2$O

Substrate - The carbon center of electrophilic attack should be sterically unhindered so that the nucleophile may donate its e$^-$ pair readily. Substrate reactivity towards Nuc: methyl > 1° > 2° > 3° (no S$_N$2)

Leaving Group - This ion, molecular fragment or molecule is replaced by the nucleophile. The most effective leaving groups are weak bases that can readily stabilize a negative charge or accept an additional e$^-$ pair. Order of leaving group ability: TsO > I > Br > Cl ~ OH$_2$ > F ~ NH$_3$ > HO, RO > NH$_2$ ~ NHR ~ NR$_2$

Solvent - The solvent plays a critical role in substitution reactions. Solvents which best sequester the positively charged counter ion raise the nucleophile's ground state energy, freeing the nucleophile to attack the substrate. Polar, aprotic solvents are best suited for this function. Solvent suitability: HMPA > CH$_3$CN > DMSO > THF > CH$_3$OH > H$_2$O

Solvents: HMPA=hexamethylphosphoramide, CH$_3$CN=acetonitrile, DMSO=dimethyl sulfoxide, THF=tetrahydrofuran.

Now that we have seen how the four factors influence the S_N2 reaction rate, let's explore the step-by-step pathway for this process.

> How S_N2 Reactions Occur

When investigation into the S_N2 reaction mechanism began more than 100 years ago, several interesting results were obtained when an optically active substrate underwent substitution: (1) a single, optically active product was obtained, and (2) the product had the opposite stereochemistry of the substrate.

This generated great interest because it ruled out reactions proceeding by carbocations since they were known to give racemic mixtures. To explain this, another mechanism must be operating. Here is an example that involves a chiral molecule, R-2-chlorobutane. We will predict the products and propose a mechanism for this transformation.

Example Problem 3. Predict the product(s) of and propose a mechanism for the reaction between KBr and R-2-chlorobutane in DMSO (dimethyl sulfoxide) solvent.

Given: Reaction cited above.
Find: (1) reaction products, (2) mechanism for the transformation.
Plan: (1) Draw all species, including transition states.
    (2) Identify nucleophile, substrate (electrophile), leaving group and solvent for the reaction.
    (3) Use the S_N2 factor guide to determine how the four factors will support an S_N2 process.
    (4) Use curved arrow formalism to show movement of electron pairs enroute to product(s).
Solve:

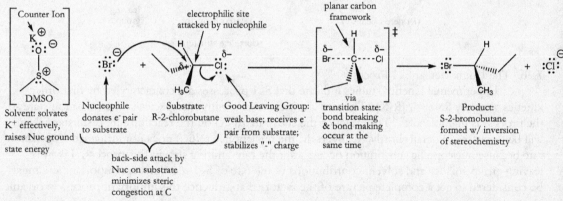

Check: (1) All species and transition state drawn correctly and identified, (2) All factors support S_N2 process, (3) bimolecular mechanism proposed consistent w/ S_N2 process. (4) Inversion of stereochemistry in product observed.

The preference for nucleophilic attack from the "backside" (opposite that of the leaving group) bears mentioning. Certainly the observation of a single product and inversion of stereochemistry is compelling evidence that the nucleophile is ready to bond to the substrate as the leaving group prepares to depart. Molecular orbital theory also provides a rationale for backside approach of the nucleophile toward the substrate. See Figure 7.2.

Br⁻   ⟶   C-Cl
HOMO   LUMO
(1 lobe)   (σ*)

Maximum overlap
achieved as Br⁻ HOMO
approaches rear σ* lobe
in a bonding fashion

Figure 7.2. Br⁻ HOMO - C-Cl LUMO Interaction.

This interaction should be familiar to you. Recall the H_2 story from chapter 2. The H-H bond in diatomic hydrogen forms as a result of a stabilizing two-electron, two orbital interaction. In Figure 7.2, the bromide HOMO with its electron pair interacts in a similar fashion with the empty C-Cl σ* molecular orbital (LUMO) 180° from the C-Cl bond. This interaction is stabilizing and results in the creation of a new Br-C bond with the observed opposite stereochemistry.

Problem 7.5. Analyze (a) by electron pushing arrows and outline pathway. Then, predict the products and propose a mechanism for (b).

(a)   :N≡C:⁻   +   [S-2-tosylbutane]   →(2 arrows, DMSO)   [R-2-methylbutanenitrile]   +   TsO:⁻ (leaving group)

(b)   :N̈=N⁺=N̈:⁻   +   [(R) stereochemistry]   →(HMPA)   [( ) stereochemistry, Complete structure]   +   _____ (leaving group)

7.3b. The Unimolecular S$_N$1 Process

    Experimental kinetics studies indicate that S$_N$1 processes are characterized by first order kinetics given by Rate = k[R-X]. The rate law is consistent with one species in the transition state so the process is unimolecular. The effect of the substrate (R-X) on the reaction rate is substantial and will be discussed in detail shortly. Conversely, the nucleophile (Nuc:) does not influence the reaction rate because nucleophilic substitution occurs after the rate-limiting step in this process. Likewise, leaving group abilities and solvent contributions to the rate of S$_N$1 reactions are important and must be considered to get a complete picture of the factors that influence this important process in organic chemistry.

    For substrates that are sterically encumbered (3°) at the carbon bearing the leaving group, nucleophilic substitution cannot occur by displacement of the leaving group from the backside. Those substrates can, however, react by a different pathway that often involves a carbocation intermediate. The component parts of this reaction are the same as the S$_N$2 process, but the four factors (nucleophile, substrate, leaving group and solvent) impact the S$_N$1 process differently than for the S$_N$2 reaction. As an example, we'll consider the substitution reaction between (*1S,2S*)–1-bromo-1-ethyl-2-methylcyclopentane and iodide in ethanol solvent at 25°C.

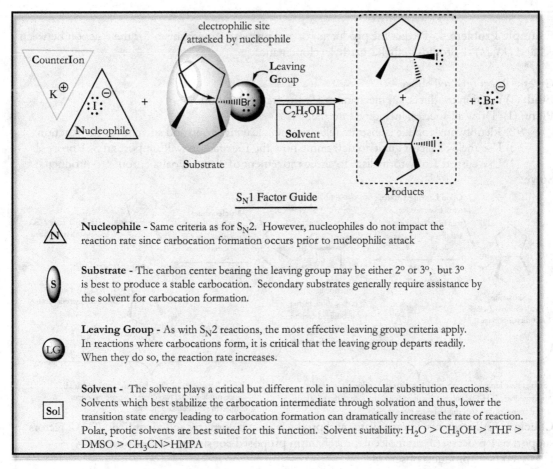

electrophilic site
attacked by nucleophile

Leaving Group

CounterIon

K⊕

Nucleophile

Substrate

C₂H₅OH

Solvent

Products

S_N1 Factor Guide

N — **Nucleophile** - Same criteria as for S_N2. However, nucleophiles do not impact the reaction rate since carbocation formation occurs prior to nucleophilic attack

S — **Substrate** - The carbon center bearing the leaving group may be either 2° or 3°, but 3° is best to produce a stable carbocation. Secondary substrates generally require assistance by the solvent for carbocation formation.

LG — **Leaving Group** - As with S_N2 reactions, the most effective leaving group criteria apply. In reactions where carbocations form, it is critical that the leaving group departs readily. When they do so, the reaction rate increases.

Sol — **Solvent** - The solvent plays a critical but different role in unimolecular substitution reactions. Solvents which best stabilize the carbocation intermediate through solvation and thus, lower the transition state energy leading to carbocation formation can dramatically increase the rate of reaction. Polar, protic solvents are best suited for this function. Solvent suitability: $H_2O > CH_3OH > THF > DMSO > CH_3CN > HMPA$

Unlike the S_N2 process that leads to inversion of stereochemistry and a single product, the S_N1 reaction often produces a mixture of stereoisomeric products. To understand why, we must now examine the mechanism of the unimolecular substitution.

➤ How S_N1 Reactions Occur

Now that we have discussed the factors which influence the S_N1 reaction, a grasp of the step-by-step process is in order. As our example, we will take a more detailed look at the reaction of (*1S,2S*)–1-bromo-1-ethyl-2-methylcyclopentane and iodide in ethanol solvent at 25°C.

Example Problem 4. Predict the product(s) of and propose a mechanism for the reaction between KI and (*1S,2S*)–1-bromo-1-ethyl-2-methylcyclopentane.

Given: Reaction cited above.
Find: (1) reaction products, (2) mechanism for the transformation.
Plan: (1) Draw all species, including intermediates.
(2) Identify nucleophile, substrate (electrophile), leaving group and solvent for the reaction.
(3) Use the S_N1 factor guide to determine how the four factors will support an S_N1 process.
(4) Use curved arrow formalism to show movement of electron pairs enroute to product(s).
Solve:

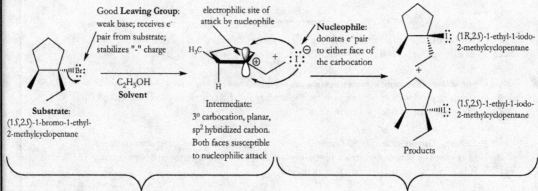

Step 1: Rate limiting ionization of 3° **substrate** produces carbocation intermediate; C_2H_5OH **solvent** assists in solvation, lowers the transition state energy of this step and stabilizes intermediate

Step 2: Rapid capture of carbocation intermediate by iodide **nucleophile**; two diastereomeric products formed in nearly equal amounts

Check: (1) All reactants, intermediates and products drawn correctly and identified, (2) All factors support S_N1 process, (3) unimolecular mechanism proposed consistent w/ S_N1 process. (4) Diastereomeric products observed.

Problem 7.6. Analyze (a) by electron pushing arrows and outline pathway. Then, predict the intermediate and leaving group and propose a mechanism for (b).

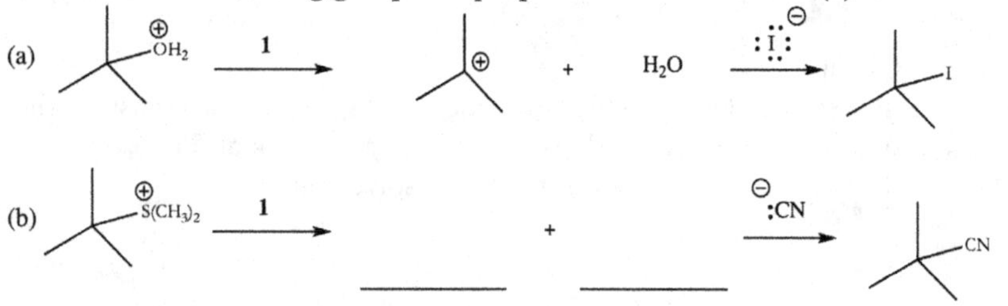

(a)

(b)

7.4 Eliminations: E2 and E1 Reactions

Elimination reactions generally follow the form shown below:

$$B{:}^- + RCH_2\text{-}CH_2X \rightarrow B\text{-}H + RCH{=}CH_2 + X{:}^-$$
$$\beta \quad \alpha$$

where one reactant, the base (B:⁻) abstracts a proton from the β-carbon adjacent to the α-carbon holding the leaving group (X is the leaving group displaced in this example). When the leaving group departs, a multiple bond is formed. We will limit our discussion in this section to two specific polar processes: elimination, bimolecular [2nd order] (E2), and elimination, unimolecular [1st order] (E1). It is important to stress here that the E2 and E1 processes we will discuss are the limiting cases along a continuum of eliminations.

78

7.4a. The Bimolecular E2 Process

Experimental kinetics studies indicate that E2 processes are characterized by second order kinetics given by Rate = k[Base][R-X]. Just as found in the S_N2 reaction, the rate law is consistent with two species in the transition state so the process is bimolecular. The effects that the base (B:) and substrate (R-X) have on the reaction rate are significant and will be discussed in detail shortly. Likewise, leaving group abilities and solvent contributions to the rate of E2 reactions are important and must be considered to get a complete picture of the factors that influence this important process in organic chemistry. Let's examine the reaction between potassium hydroxide and R-2-bromobutane below:

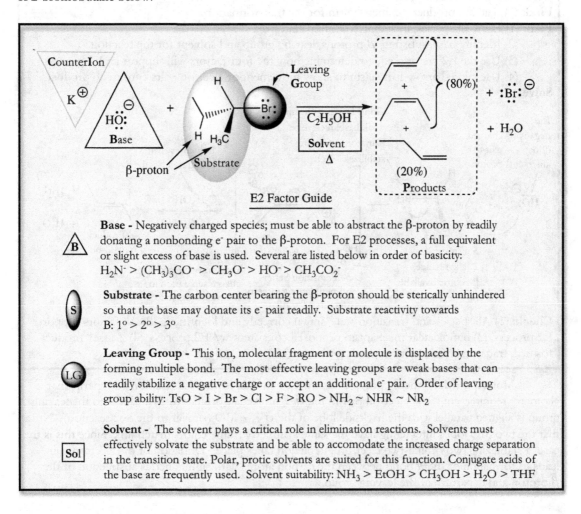

E2 Factor Guide

Base - Negatively charged species; must be able to abstract the β-proton by readily donating a nonbonding e⁻ pair to the β-proton. For E2 processes, a full equivalent or slight excess of base is used. Several are listed below in order of basicity: $H_2N^- > (CH_3)_3CO^- > CH_3O^- > HO^- > CH_3CO_2^-$

Substrate - The carbon center bearing the β-proton should be sterically unhindered so that the base may donate its e⁻ pair readily. Substrate reactivity towards B: 1° > 2° > 3°

Leaving Group - This ion, molecular fragment or molecule is displaced by the forming multiple bond. The most effective leaving groups are weak bases that can readily stabilize a negative charge or accept an additional e⁻ pair. Order of leaving group ability: $TsO > I > Br > Cl > F > RO > NH_2 \sim NHR \sim NR_2$

Solvent - The solvent plays a critical role in elimination reactions. Solvents must effectively solvate the substrate and be able to accomodate the increased charge separation in the transition state. Polar, protic solvents are suited for this function. Conjugate acids of the base are frequently used. Solvent suitability: $NH_3 > EtOH > CH_3OH > H_2O > THF$

Reaction conditions are important here. The use of at least one equivalent of a strong base makes it unlikely for this reaction to proceed by any mechanism other than an E2 process. Running the reaction at elevated temperature minimizes the likelihood of the base functioning as a nucleophile resulting in an S_N2 process.

The products of this reaction, trans-2-butene, cis-2-butene and 1-butene, also bear mentioning. Product distribution in elimination reactions is governed by **Zaitsev's Rule**, namely that the more highly substituted alkenes are the major products formed. Recall from Chapter 4 that since disubstituted alkenes are more stable than monosubstituted alkenes, the 2-butenes comprise the majority products as expected. Of the 2-butene products, trans-2-butene is the major product (~45% of the total product mixture) since trans alkenes are more stable with less steric interference than cis alkenes.

➢ How E2 reactions occur

Now that we have discussed the factors which influence the E2 reaction, a grasp of the step-by-step process is in order. The example problem below gives a more detailed look at an E2 reaction mechanism.

Example Problem 5. Propose a mechanism for the reaction between *R*–2-bromobutane and KOH in ethanol solvent at 50°C to produce trans-2-butene.

Given: Reaction cited above.
Find: (1) reaction product, (2) mechanism for the transformation.
Plan: (1) Draw all species, including transition states.
       (2) Identify base, substrate β-protons, leaving group and solvent for the reaction.
       (3) Use the E2 factor guide to determine how the four factors will support an E2 process.
       (4) Use curved arrow formalism to show movement of electron pairs enroute to product.
Solve:

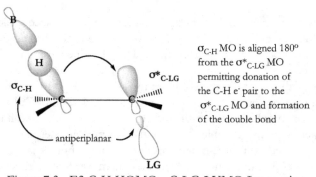

Check: (1) All species and transition state drawn correctly and identified, (2) All factors support E2 process, (3) bimolecular mechanism proposed consistent w/ E2 process. (4) Zaitsev product formed; trans stereochemistry in product observed.

Molecular orbital theory provides us with a rationale for the E2 reaction's antiperiplanar geometry requirement. The electron pair in the $\sigma_{C\text{-}H}$ molecular orbital (MO) adjacent to the leaving group is aligned parallel with the backside lobe of the $\sigma^*_{C\text{-}LG}$ MO (similar to the S_N2 reaction). Note that the two MO lobes interacting have the same symmetry (same colors match up). Since this is the case, a stabilizing two orbital-two electron interaction takes place. Donation of the electron pair leads to formation of the π bond between the carbon atoms with simultaneous abstraction of the β-proton and departure of the leaving group.

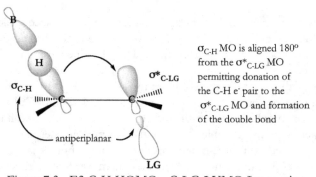

$\sigma_{C\text{-}H}$ MO is aligned 180° from the $\sigma^*_{C\text{-}LG}$ MO permitting donation of the C-H e⁻ pair to the $\sigma^*_{C\text{-}LG}$ MO and formation of the double bond

Figure 7.3. E2 C-H HOMO - C-LG LUMO Interaction.

> An exception to Zaitsev's rule: Hoffman eliminations

There are eliminations which do not give, as the major products, the more highly substituted alkene products in accord with Zaitsev's rule. These non-Zaitsev or Hoffman eliminations are likely to occur when one or more of the following apply: (1) the presence of a poor, sterically hindered or bulky leaving group and/or (2) the use of a very bulky base. As an example, consider the reaction between *S*-2-trimethylammoniumpentane and tert-butoxide to yield 1-pentene:

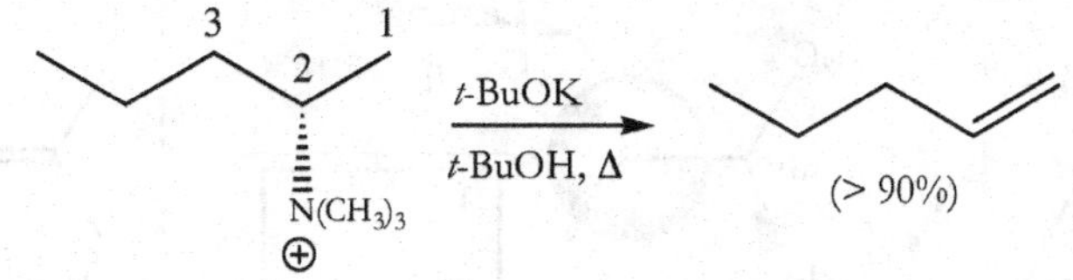

If we view the Newman projections of the *S*-2-trimethylammoniumpentane starting material, we can readily discern why the Hoffman elimination product is favored:

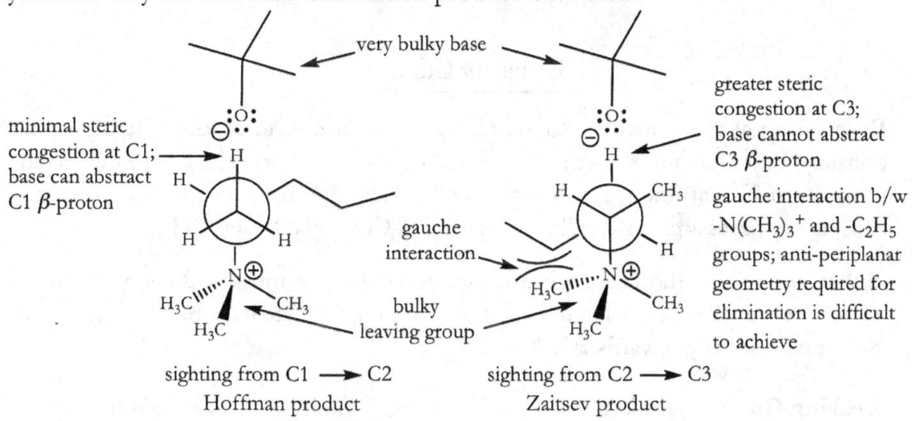

Problem 7.7. Analyze (a) by electron pushing arrows and outline pathway. Then, predict the products and propose a mechanism for (b).

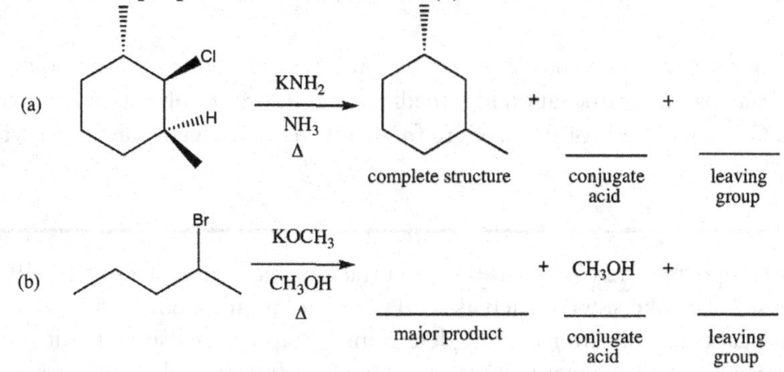

7.4b. The Unimolecular E1 Process

Just like the S_N1 reaction we examined earlier, experimental kinetics studies indicate that E1 processes are characterized by first order kinetics given by Rate = k[R-X]. The rate law is consistent with one species in the transition state so the process is unimolecular. The effects of the substrate (R-X), leaving group and solvent on the reaction rate are substantial and will be discussed in detail shortly. Conversely, the base (B:) does not influence the reaction rate because abstraction of the β-proton occurs after the rate-limiting step in this process. The interplay of these factors on the rate of E1 reactions is important and must be considered to get a complete picture of how they influence this important process in organic chemistry.

For substrates that are sterically encumbered (3°) at the carbon bearing the leaving group, the initial step in the E1 pathway involves a carbocation intermediate - the same as for the S_N1 reaction. While the component parts of this reaction are the same as the E2 process, the four factors (base, substrate, leaving group and solvent) impact the E1 process differently than for the E2 reaction. As an example, we'll consider the elimination reaction between 2-bromo-2,2-dimethylpropane and ethoxide (5 mol% base:substrate) in ethanol solvent at 50°C.

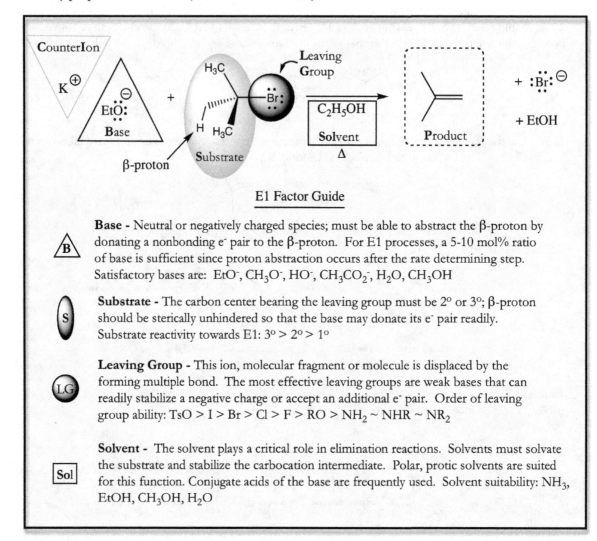

E1 Factor Guide

Base - Neutral or negatively charged species; must be able to abstract the β-proton by donating a nonbonding e⁻ pair to the β-proton. For E1 processes, a 5-10 mol% ratio of base is sufficient since proton abstraction occurs after the rate determining step. Satisfactory bases are: EtO^-, CH_3O^-, HO^-, $CH_3CO_2^-$, H_2O, CH_3OH

Substrate - The carbon center bearing the leaving group must be 2° or 3°; β-proton should be sterically unhindered so that the base may donate its e⁻ pair readily. Substrate reactivity towards E1: 3° > 2° > 1°

Leaving Group - This ion, molecular fragment or molecule is displaced by the forming multiple bond. The most effective leaving groups are weak bases that can readily stabilize a negative charge or accept an additional e⁻ pair. Order of leaving group ability: $TsO > I > Br > Cl > F > RO > NH_2 \sim NHR \sim NR_2$

Solvent - The solvent plays a critical role in elimination reactions. Solvents must solvate the substrate and stabilize the carbocation intermediate. Polar, protic solvents are suited for this function. Conjugate acids of the base are frequently used. Solvent suitability: NH_3, $EtOH$, CH_3OH, H_2O

Reaction conditions are important here. The use of a neutral base or a small quantity (5-10 mol% equivalent) of anionic base/alcoholic solvent such as KOH/ethanol is sufficient to carry out the E1 elimination step since it occurs rapidly after the rate determining step. This also makes it unlikely for this reaction to proceed by an E2 process since that requires a full equivalent of strong base. Running the reaction at elevated temperature minimizes the likelihood of the base functioning as a nucleophile resulting in an S_N1 process.

The single product of this reaction is 2-methylpropene. Like product distribution in E2 reactions, the products of E1 reactions with more complicated substrates than the one just discussed are governed by **Zaitsev's Rule**.

➢ How E1 reactions occur

Now that we have discussed the factors which influence the E1 reaction, a grasp of the step-by-step process is in order. The example problem below gives a more detailed look at an E1 reaction mechanism.

Example Problem 6. Propose a mechanism for the reaction between ethoxide (5 mol%) and 2-bromo-2,2-dimethylpropane in ethanol solvent at 50°C to produce 2-methylpropene.

Given: Reaction cited above.

Find: (1) reaction product, (2) mechanism for the transformation.

Plan: (1) Draw all species, including intermediates.
   (2) Identify base, substrate β-protons, leaving group and solvent for the reaction.
   (3) Use the E1 factor guide to determine how the four factors will support an E1 process.
   (4) Use curved arrow formalism to show movement of electron pairs enroute to product.

Solve:

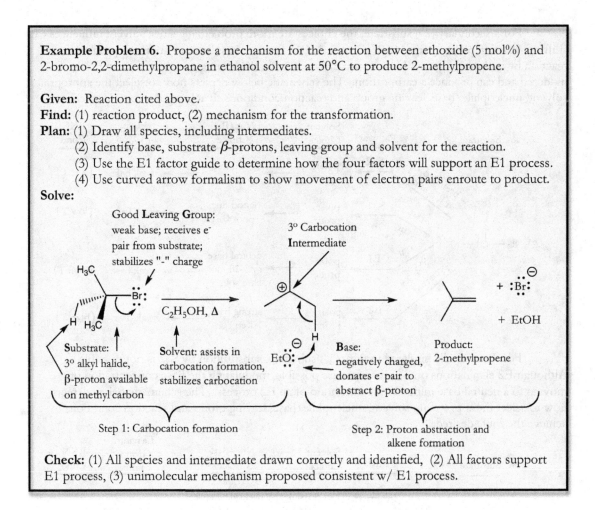

Check: (1) All species and intermediate drawn correctly and identified, (2) All factors support E1 process, (3) unimolecular mechanism proposed consistent w/ E1 process.

7.5 Competition between Nucleophilic Substitutions and Eliminations

We have now looked at the four limiting cases - S_N2, S_N1, E2 and E1 and examined how the four factors - Nucleophile/Base, Substrate, Leaving Group and Solvent impact these processes. For methyl substrates, only S_N2 substitutions can occur since S_N1 would require formation of the highly unstable methyl carbocation. However, for 1°, 2° and 3° substrates, two or more pathways frequently compete with each other during the course of a reaction. Optimizing a reaction to proceed by the nucleophilic substitution or elimination pathway of our choosing requires that we carefully consider each factor and select the best possible conditions to ensure success.

For a primary (1°) substrate (R≠H), the schematic below shows how to select the appropriate solvent, nucleophile/base, leaving group and reaction conditions to achieve the route desired:

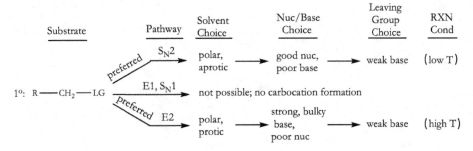

For a secondary (2°) substrate, the choices are more problematic since solvent influences can shift the reaction pathway from S_N2/E2 to S_N1/E1. Secondary substrates are more prone to reaction by multiple pathways because the carbon bearing the leaving group is more sterically hindered and can produce a carbocation. The schematic below depicts how to select the appropriate solvent, nucleophile/base, leaving group and reaction conditions to achieve the route desired:

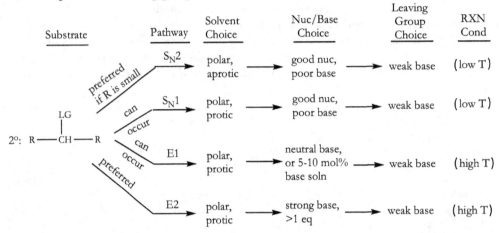

For the sterically hindered tertiary (3°) substrate, substitution by S_N2 is not possible. Although E2 eliminations on 3° substrates are possible, modulating the base concentration or moving to a neutral base mitigates the likelihood of an E2 process. The schematic below depicts how to select the appropriate solvent, nucleophile/base, leaving group and reaction conditions to achieve the route desired:

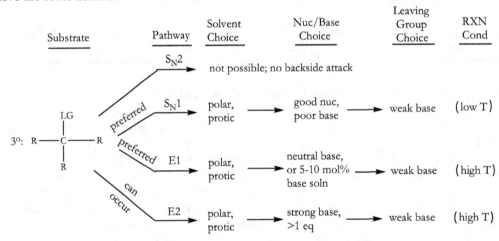

Problem 7.8. Analyze (a) by electron pushing arrows and outline pathway. Then, predict the products and propose a mechanism for (b).

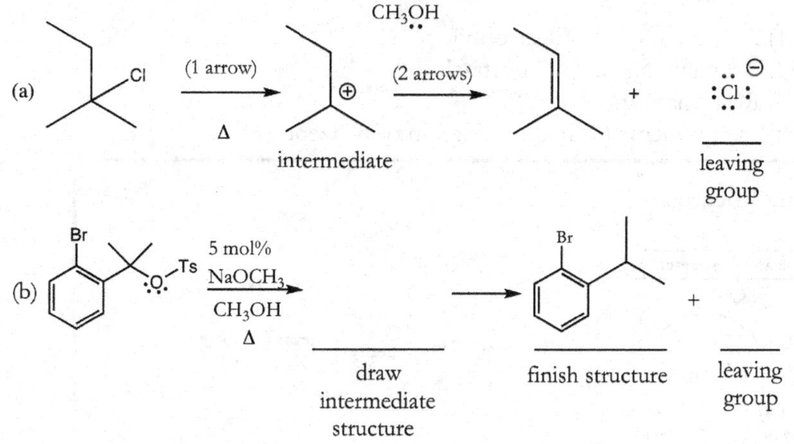

(a)

(1 arrow) Δ → intermediate → (2 arrows) CH₃ÖH → + :Cl̈:⊖ leaving group

(b)

5 mol% NaOCH₃ CH₃OH Δ → draw intermediate structure → finish structure + leaving group

7.6 Electrophilic Aromatic Substitution (EAS)

In Chapter 4, the unique stability of benzene was discussed. In fact, when chemists in the 19th century began to explore benzene's reactivity towards electrophiles, they found that this aromatic molecule failed to undergo addition reactions typical of other alkenes. For example, Figure 7.4 shows that while cyclohexene undergoes electrophilic addition of Br_2, to yield trans-1,2-dibromocyclohexane, benzene does not yield a dibromo product. If, however, benzene was allowed to react with Br_2 in the presence of a Lewis Acid such as Fe, a different monobrominated product resulted, bromobenzene.

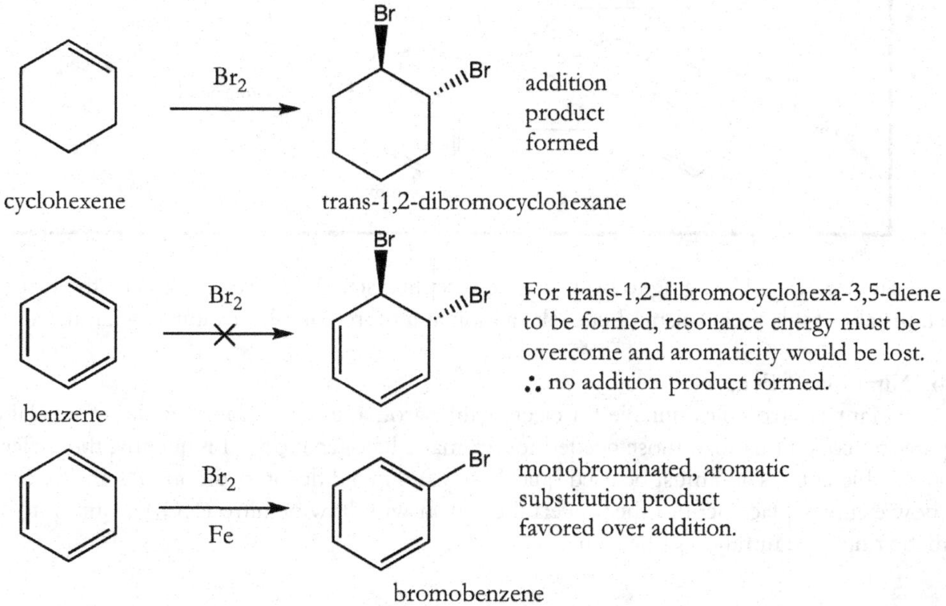

cyclohexene → trans-1,2-dibromocyclohexane   addition product formed

benzene → ✗   For trans-1,2-dibromocyclohexa-3,5-diene to be formed, resonance energy must be overcome and aromaticity would be lost. ∴ no addition product formed.

benzene → bromobenzene   monobrominated, aromatic substitution product favored over addition.

Figure 7.4. Bromination of Cyclohexene and Benzene.

The bromine had replaced one of the protons on the benzene molecule and the product was aromatic, so that reaction was a type of substitution. However, it was found that benzene reacted with most electrophiles only in the presence of Lewis acid catalysts. To understand why this is the case, let's examine some representative EAS mechanisms a little more closely.

7.6.a. Bromination of Benzene

It turns out that most electrophilic aromatic substitutions can be thought of as having three distinct components:

(1) Activation of the Electrophile
(2) Aromatic Nucleophilic Attack
(3) Rearomatization

Let's apply these mechanistic components to the bromination of benzene:

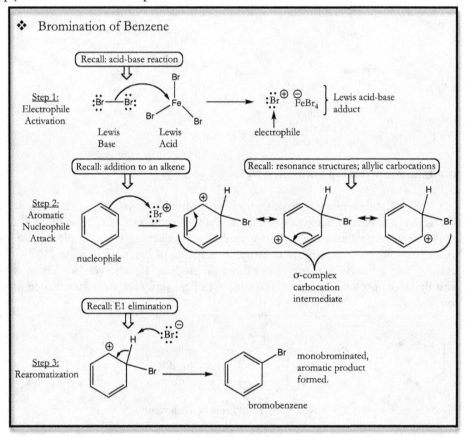

As you can see, we are bringing to bear many concepts discussed in earlier parts of the text to answer questions about this mechanism. Let's take a look at a more complex example – nitration.

7.6.b. Nitration of Benzene

Many electrophiles suitable for electrophilic aromatic substitution require more elaborate preparation conditions than those needed for aromatic halogenations. Frequently, these electrophiles are not stable entities and must be used quickly so as to avoid decomposition. As an example, we will now examine placement of the powerful electron-withdrawing nitro (NO_2) group onto an aromatic ring - the nitration of benzene.

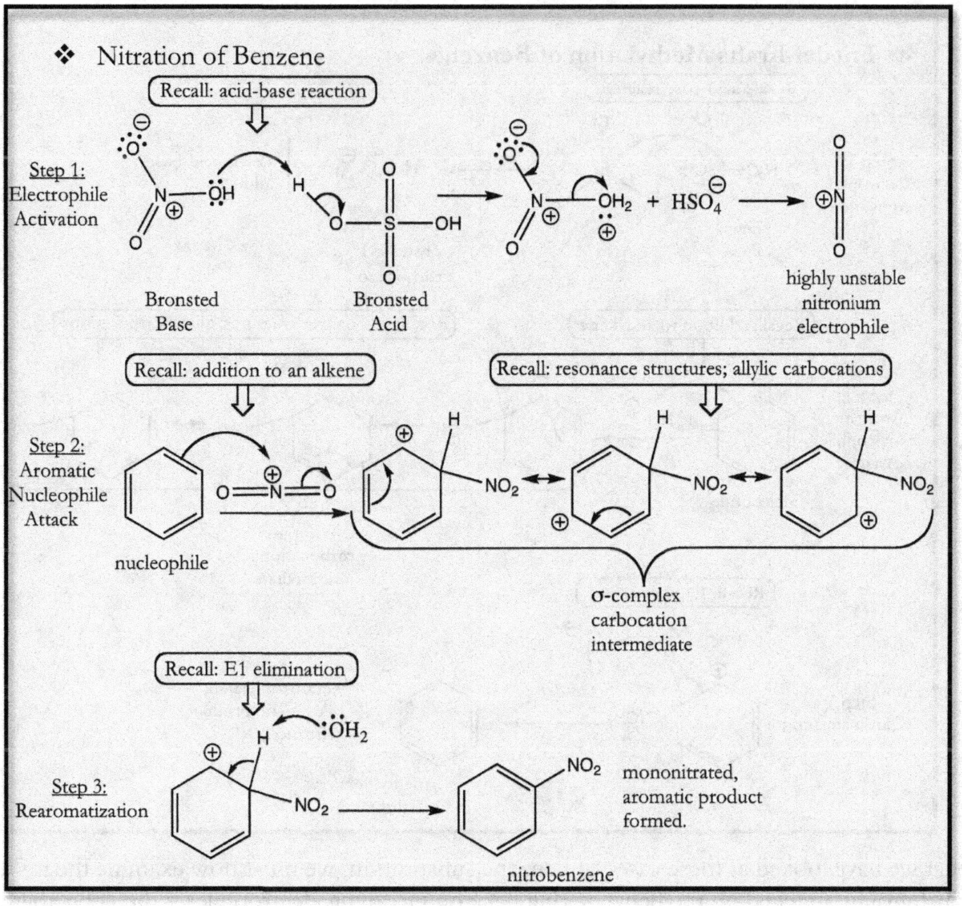

❖ Nitration of Benzene

Recall: acid-base reaction

Step 1: Electrophile Activation

Bronsted Base

Bronsted Acid

+ HSO₄⁻

highly unstable nitronium electrophile

Recall: addition to an alkene

Recall: resonance structures; allylic carbocations

Step 2: Aromatic Nucleophile Attack

nucleophile

σ-complex carbocation intermediate

Recall: E1 elimination

Step 3: Rearomatization

mononitrated, aromatic product formed.

nitrobenzene

An important point to be made here is the nature of the nitration σ-complex. Resonance structures can be drawn which place positive formal charges next to the powerful NO_2 EWG. The nitro group is also an EWG by induction. These combined effects destabilize the σ-complex and that has implications for nitrobenzene's reactivity with additional electrophiles. We will address those shortly. Before we do though, let's examine the Friedel-Krafts alkylation - placement of an electron-donating alkyl group onto the benzene ring.

7.6.c. Friedel-Krafts Alkylation of Benzene

Our final mechanistic study involves placement of a methyl group onto a benzene ring. In this process, the electrophile generated is a methyl carbocation, which we know from Chapter 4 to be unstable. As with the nitronium ion electrophile discussed earlier, the methyl carbocation is generated in the same reaction vessel (*in situ*) as is the benzene. Upon production of the electrophile, the benzene immediately reacts with it, giving the electrophile no opportunity to decompose.

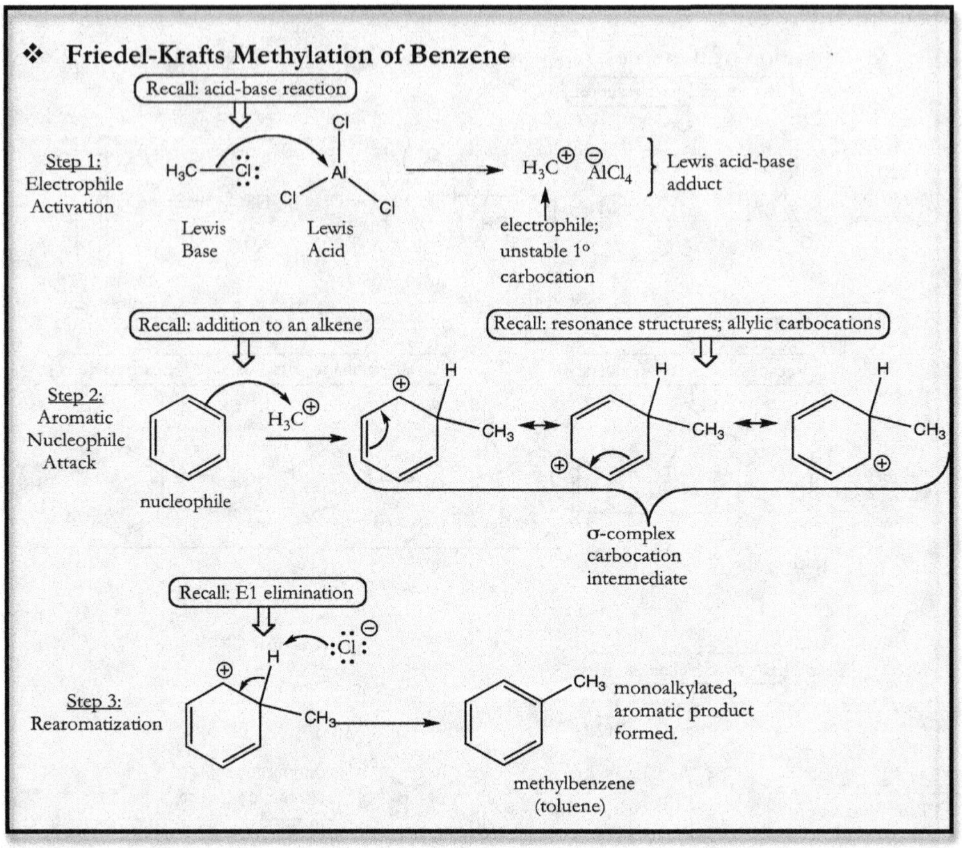

Friedel-Krafts Methylation of Benzene

Now that we have looked at these cases of benzene substitution, we must now examine the influence functional groups attached on the benzene ring have on incoming electrophiles – directing ability.

7.6.d. Directing Ability of Substituents on the Benzene Ring

Once a substituent is attached to the benzene ring, it controls or directs further electrophilic aromatic substitutions on the ring. Functional groups on the benzene ring can direct incoming electrophiles to certain carbon positions on the ring – *ortho* (*o*), *meta* (*m*) or *para* (*p*) to the attached group. What causes some functional groups to direct to one of these positions is the ability to activate that position by increasing electron density in the ring. Other functional groups actually deactivate the ring toward further EAS by depleting electron density in the ring. As it happens, most substituents that activate the ring are *o – p* directors, while those that deactivate the ring are usually *m* directors. Let's explore why.

❖ Electron Donors: *o – p* Directors

Electron donation to the benzene ring occurs by both hyperconjugation and resonance (π system) pathways. Alkyl groups donate electron density by hyperconjugation while substituents with nonbonding electron pairs on the atoms next to the benzene ring donate by resonance. To illustrate the paths of electron donation, let's consider a donor's (D) effect on the *ortho*, *meta* and *para* substituted σ complex carbocation intermediates:

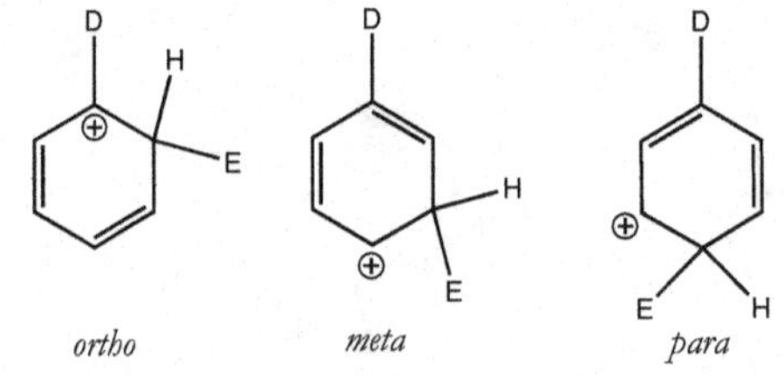

ortho   *meta*   *para*

We will use the methyl and amino[3] electron donating groups (EDGs) as our donors:

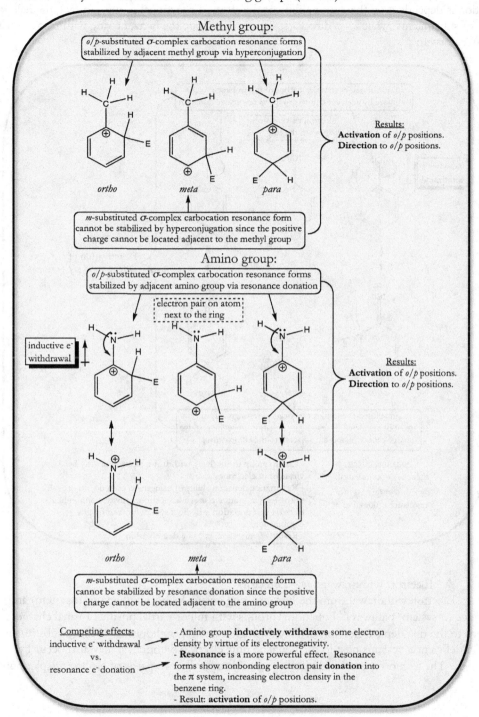

As noted in the amino group example, opposing resonance and inductive effects compete. Yet, the amino group is a very most powerful activating group. In fact, aniline undergoes tribromination at room temperature with no catalyst to yield 2,4,6-tribromoaniline. If the temperature is lowered to 0°C, 2-bromoaniline is obtained.[3]

[3]Shultz, D. A., Sloop, J.C., Washington, G.E. *J. Org. Chem.*, **2006**, *71*, 9104.

While any atom with nonbonding electron pairs may be an $o - p$ director, the degree of ring activation is dependent on the donor ability of the atom or functional group. In fact, the halogen family of substituents, while $o - p$ directors, slightly deactivate the benzene ring with respect to attack on benzene itself:

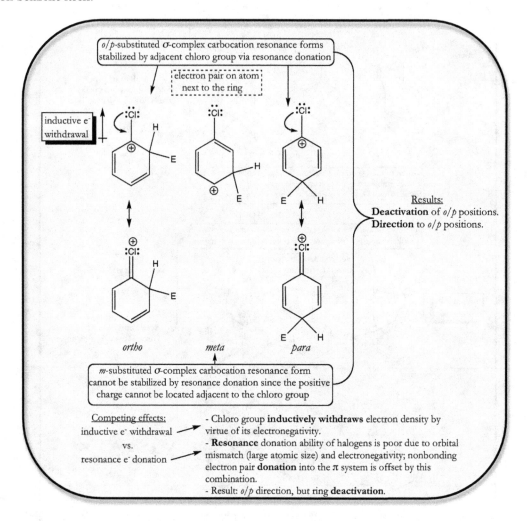

- ❖ Electron Withdrawers: m – Directors

Electron withdrawal from the benzene ring occurs by both inductive (σ system) and resonance (π system) pathways. Functional groups with full or partial positive formal charges adjacent to the ring deplete electron density through inductive withdrawal. If those substituents have electron deficient π systems next to the benzene ring, electron withdrawal may also occur by resonance. The trifluoromethyl and nitro groups illustrate the two paths of electron depletion:

Trifluoromethyl group:

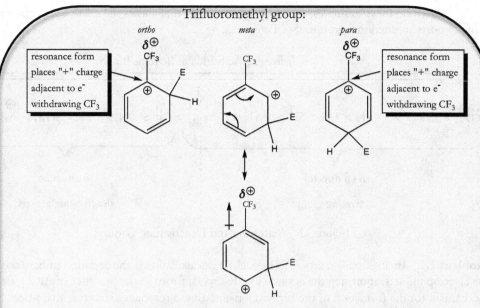

- Trifluoromethyl group **inductively withdraws** electron density by virtue of the δ^+ on the CF_3 carbon due to fluorine's electronegativity. This destabilizes the σ complex carbocation.
- **Resonance** forms for the *o/p*-substituted σ complex carbocation destabilized substantially by presence of a "+" charge adjacent to the trifluoromethyl group; least stable intermediates
- **Resonance** forms for the *m*-substituted σ complex carbocation place a "+" charge further away from the trifluoromethyl group; more stable intermediates
- Result: **deactivation** of *all* positions, but *meta* position is affected least; meta director.

Nitro group:

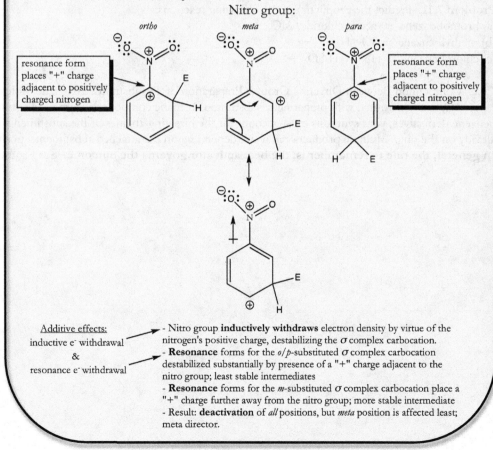

Additive effects:
inductive e⁻ withdrawal
&
resonance e⁻ withdrawal

- Nitro group **inductively withdraws** electron density by virtue of the nitrogen's positive charge, destabilizing the σ complex carbocation.
- **Resonance** forms for the *o/p*-substituted σ complex carbocation destabilized substantially by presence of a "+" charge adjacent to the nitro group; least stable intermediates
- **Resonance** forms for the *m*-substituted σ complex carbocation place a "+" charge further away from the nitro group; more stable intermediate
- Result: **deactivation** of *all* positions, but *meta* position is affected least; meta director.

Based on these examples, we may prepare a chart that shows the relative activating ability of some important functional groups. See Figure 7.5.

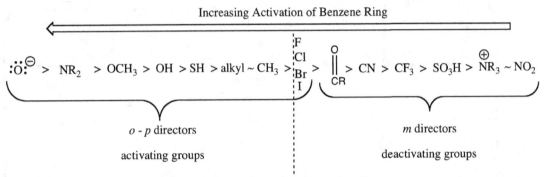

Figure 7.5. Activating and Deactivating Groups.

Problem 7.9. An analogous reaction to Friedel-Krafts alkylation is the acylation of benzene. During the electrophile activation step the acylium ion, $R-C\equiv O^+$, forms in the presence of $AlCl_3$. Propose a mechanism for the formation of the acylium ion and draw a resonance structure that places the positive formal charge on carbon.

Problem 7.10. Benzene rings are sulfonated using a SO_3/H_2SO_4 mixture. During the electrophile activation step, a protonated sulfur trioxide ion, $^+SO_3H$, is formed. Propose a mechanism for the formation of the $^+SO_3H$ ion and pathway for the sulfonation of benzene.

Problem 7.11. Predict the product(s) of the following reactions:
(a) bromobenzene + acetyl chloride/$AlCl_3$ →
(b) methylbenzene + SO_3/H_2SO_4 →
(c) chlorobenzene + HNO_3/H_2SO_4 →

7.6.e. Competition Between Directing Groups: Preparation of Trisubstituted Aromatic Molecules
     When confronted with preparing trisubstituted aromatic compounds from disubstituted benzene derivatives, your synthesis must account for the directing abilities of the substituents that are already on the ring. Multiple products can result depending on the attached substituents' positions. **In general, the rule to remember is: the best activator governs the outcome.** See Figure 7.6.

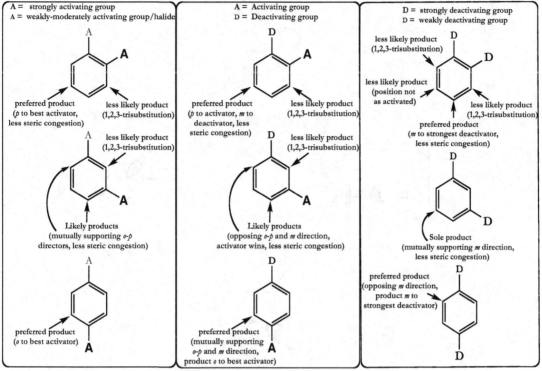

Figure 7.6. Electrophilic Aromatic Substitution Patterns for Disubstituted Arenes.

Problem 7.12. Predict the product(s) of the following reactions:
(a) 1-bromo-2-methylbenzene (2-bromotoluene) + acetyl chloride/AlCl$_3$ →
(b) 1-methoxy-3-nitrobenzene (3-nitroanisole) + SO$_3$/H$_2$SO$_4$ →
(c) 4-chlorobenzene sulfonic acid + HNO$_3$/H$_2$SO$_4$ →

7.7 Nucleophilic Addition to Carbonyl Species and Nucleophilic Acyl Substitution

As we learned in Chapter 6, the carbonyl functional group is an electrophilic species owing to its strong bond dipole:

$$\Delta E_n \text{ (C-O)} = 1, \therefore \text{large bond dipole.}$$

The carbon, which bears a δ^+ charge, is subject to attack by nucleophilic species.

This process takes place with very specific geometric and steric considerations; we will address those shortly. Why nucleophilic addition and nucleophilc acyl substitution to carbonyls takes place is best explained by MO theory. See Figure 7.7. In this case, we consider the two-orbital, two-electron interaction between the HOMO of the nucleophile and the LUMO of the carbonyl:

93

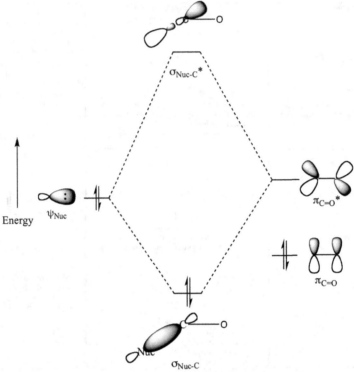

Figure 7.7. Nuc⁻ HOMO – C=O LUMO Interaction.

Notice that because the filled Nuc HOMO (Ψ_{Nuc}) and the unfilled C=O LUMO ($\pi_{C=O}$*) are close in energy we expect this interaction to be stabilizing and strong. Hence, HOMO-LUMO overlap creates a substantial energy savings upon formation of the σ_{Nuc-C} bond. Additionally, the fact that the C=O HOMO ($\pi_{C=O}$) is filled would create a destabilizing, two orbital-four electron interaction were it to combine with the Nuc HOMO.

7.7.a. Reactivity of Carbonyl Species Towards Nucleophilic Attack

There are numerous functional groups that contain carbonyls. Structurally, they can be thought of in terms of carbonyls that contain no leaving groups – aldehydes and ketones, and the carboxylic acid derivative family of compounds (those that possess a leaving group) - acid halides, anhydrides, esters, and amides. Reactivity of these species is dependent upon both steric congestion at the carbonyl carbon and the degree of partial positive charge character the carbonyl carbon possesses. The graphic (Figure 7.8) explains the trend in reactivity.

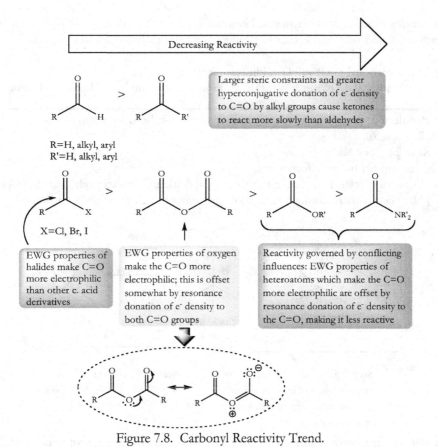

Decreasing Reactivity

R=H, alkyl, aryl
R'=H, alkyl, aryl

Larger steric constraints and greater hyperconjugative donation of e⁻ density to C=O by alkyl groups cause ketones to react more slowly than aldehydes

X=Cl, Br, I

EWG properties of halides make C=O more electrophilic than other c. acid derivatives

EWG properties of oxygen make the C=O more electrophilic; this is offset somewhat by resonance donation of e⁻ density to both C=O groups

Reactivity governed by conflicting influences: EWG properties of heteroatoms which make the C=O more electrophilic are offset by resonance donation of e⁻ density to the C=O, making it less reactive

Figure 7.8. Carbonyl Reactivity Trend.

7.7.b. Nucleophilic Addition and Acyl Substitution to Carbonyls
General overall reactions for these processes are shown below:

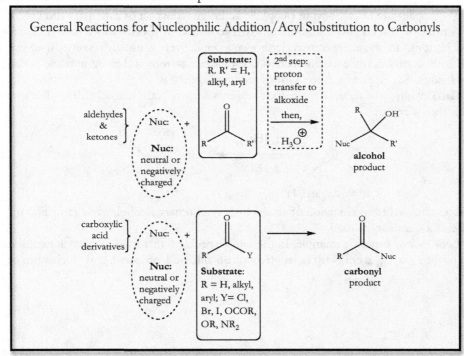

General Reactions for Nucleophilic Addition/Acyl Substitution to Carbonyls

aldehydes & ketones

Nuc:
neutral or negatively charged

Substrate:
R. R' = H, alkyl, aryl

2nd step: proton transfer to alkoxide then, $H_3O^{\oplus}$

alcohol product

carboxylic acid derivatives

Nuc:
neutral or negatively charged

Substrate:
R = H, alkyl, aryl; Y= Cl, Br, I, OCOR, OR, NR₂

carbonyl product

95

The lack of leaving groups for aldehydes and ketones results in 1,2-addition alcohol products whereas carboxylic acid derivatives react to produce different carbonyl species.

➤ How carbonyl nucleophilic additions occur
Let's now take a closer look at the mechanism for simple aldehydes and ketones:

Example Problem 7. Propose a mechanism for the reaction between methyllithium and ethanal to produce 2-propanol.
Given: Reaction cited above.
Find: Mechanism for the transformation.
Plan: (1) Draw all species, including intermediates. (2) Identify nucleophile and electrophile
(3) Use curved arrow formalism to show movement of electron pairs enroute to product.
Solve:

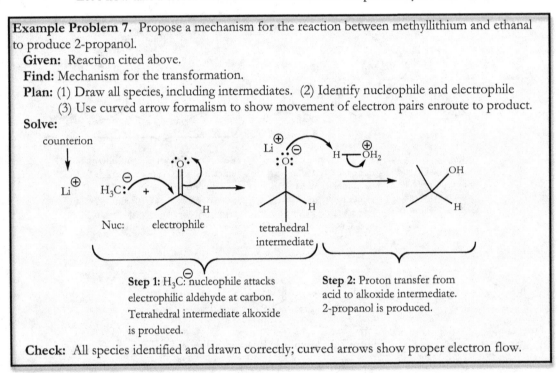

Step 1: $H_3C:^{\ominus}$ nucleophile attacks electrophilic aldehyde at carbon. Tetrahedral intermediate alkoxide is produced.

Step 2: Proton transfer from acid to alkoxide intermediate. 2-propanol is produced.

Check: All species identified and drawn correctly; curved arrows show proper electron flow.

The net result is a 1,2-addition across the C=O.

➤ Application of carbonyl nucleophilic addition: reduction of aldehydes and ketones
A principle reaction in organic chemistry is reduction by species which deliver electrons, H atoms, or H:⁻ ions. In organic chemistry, hydrogen atom delivery is usually accomplished via metal catalyzed hydrogenation, while metals such as zinc deliver electrons and aluminum and boron provide H:⁻ ions. See Appendix A for a list of reducing agents.

Carbonyls may be reduced by hydride reagents such as $NaBH_4$ and $LiAlH_4$. The general reaction is shown below:

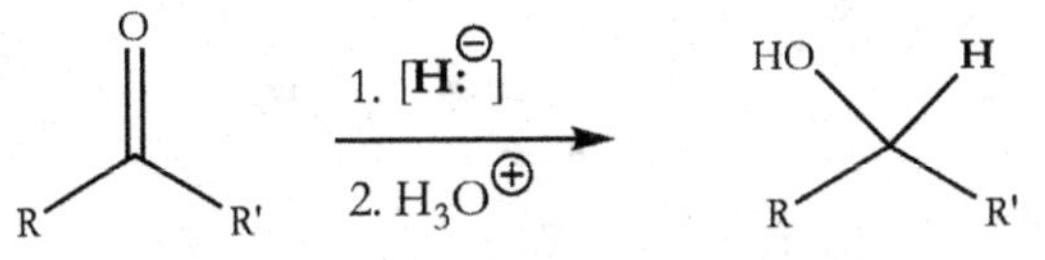

$$\underset{R \quad\quad R'}{\overset{O}{\big\|}} \quad \xrightarrow[\text{2. } H_3O^{\oplus}]{\text{1. } [\mathbf{H:}^{\ominus}]} \quad \underset{R \quad\quad R'}{\overset{HO \quad\quad H}{\diagdown\diagup}}$$

R, R' = alkyl aryl, H

The product obtained from reduction of an aldehyde is a primary alcohol, while reduction of a ketone yields a secondary alcohol.

Let's look at a specific example. In the sample problem that follows, we will examine the function of the reducing agent $NaBH_4$, its effect on an aldehyde, propanal, and mechanism of reduction.

Example problem 8. Predict the reduction product of the following reaction and propose a mechanism for the reaction.

$$CH_3CH_2CHO \xrightarrow[\text{2. } H_3O^+]{\text{1. } NaBH_4} \quad \textbf{?}$$

Given: CH_3CH_2CHO and   1. $NaBH_4$, 2. H_3O^+

Find: Reduction product, reaction mechanism

Plan: 1. Draw reactant structures
   2. Identify reactive center(s) in the substrate
   3. Identify reagent function based on structure

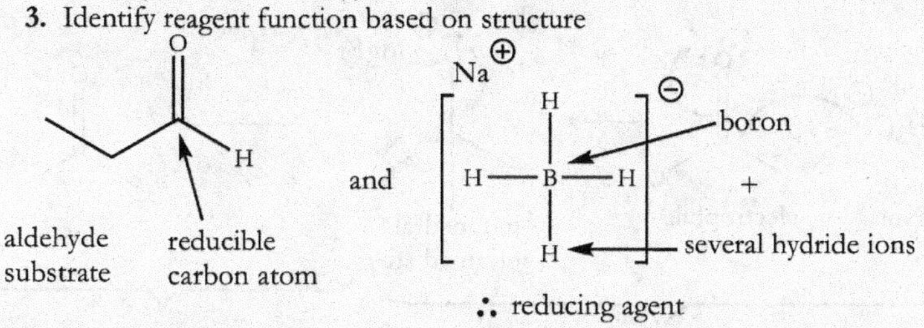

aldehyde   reducible   and   ∴ reducing agent
substrate   carbon atom

   4. Propose a mechanism

Solve:

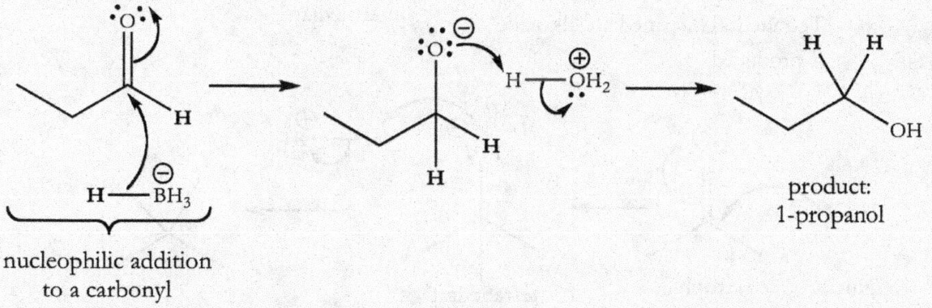

nucleophilic addition
to a carbonyl

product:
1-propanol

Check: 1. Reactant structures drawn correctly, 2. Reactive center identified in the substrate, 3. Reagent function identified based on structure, 4. Mechanism proposed using curved arrow formalism consistent with reduction of aldehydes by hydride reagents.

➤ How nucleophilic acyl substitutions occur

Here's a look at the mechanism for the reaction between the Grignard reagent, methylmagnesium bromide and acetyl chloride(a carboxylic acid derivative) to produce 2-methyl-2-propanol.

Example Problem 9. Propose a mechanism for the reaction between methyllithium and acetyl chloride to produce 2-methyl-2-propanol.

  Given: Reaction cited above.

  Find: Mechanism for the transformation.

  Plan: (1) Draw all species, including intermediates.

  (2) Identify nucleophile, electrophile and leaving group.

  (3) Use curved arrow formalism to show movement of electron pairs enroute to product.

Solve:

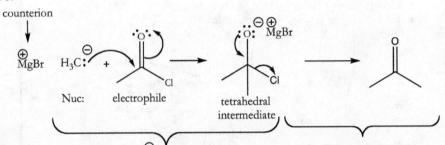

Step 1: H₃C: nucleophile attacks electrophilic acid chloride at carbon. Tetrahedral intermediate alkoxide is produced.

Step 2: Tetrahedral intermediate collapses, producing acetone intermediate.

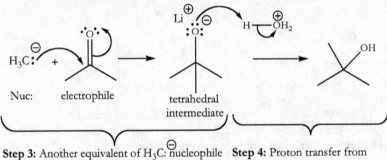

Step 3: Another equivalent of H₃C: nucleophile attacks electrophilic ketone at carbon. Tetrahedral intermediate alkoxide is produced.

Step 4: Proton transfer from acid to alkoxide intermediate. 2-methyl-2-propanol is produced.

Check: All species identified and drawn correctly; curved arrows show proper electron flow. One note in particular bears mentioning at this point. In cases of carboxylic acid derivatives that have poor leaving groups (esters and amides), protonation of the alkoxide and amide leaving groups may be required to effect ejection of the leaving species.

The net result is a 1,2-addition across the C=O.

> ➤ Stereoelectronic considerations

For the simple C=O examples we have just examined, the incoming nucleophile may approach either face of the $\pi_{C=O}$* MO at the Bürgi-Dunitz angle (109°). For aldehyde and ketone substrates, a mixture of enantiomeric alcohols can result. In some cases, the mixture is racemic.

When, however, the substrate α-carbon (carbon next to the carbonyl) contains a chiral center and/or heteroatoms capable of chelating the Lewis acid counter ion of the nucleophile, predicting the nucleophile's trajectory toward the carbonyl LUMO requires special consideration. The Felkien-Ahn model is often used in to explain the nucleophile's approach path in carbonyls that are chiral at the α-carbon. The Chelation model enables us to rationalize the nucleophilic approach path in carbonyls that bear heteroatoms capable of chelating the Lewis acid counter ions of the nucleophile. These models also allow us to successfully predict the diastereotopic products that result from nucleophilic addition to these carbonyls. See Figure 7.9.

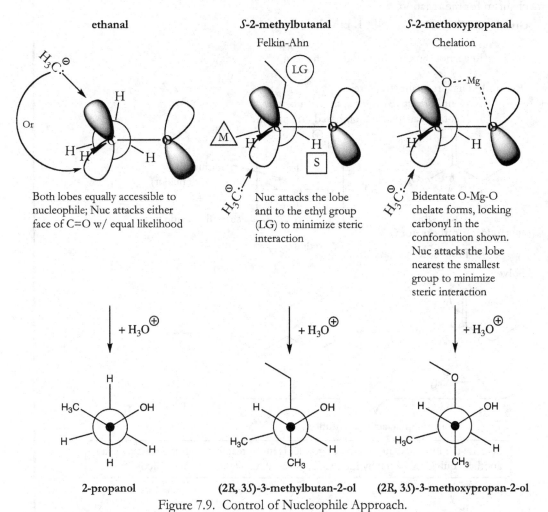

Figure 7.9. Control of Nucleophile Approach.

7.7. c. Reduction of Esters: A Double-Barrel Reaction: Nucleophilic Acyl Substitution and Nucleophilic Addition to a Carbonyl

As we learned earlier, hydride reagents reduce aldehydes and ketones. We will now examine the effect of a hydride reducing agent on an ester (a carboxylic acid derivative). The general reaction is:

99

In the example that follows, we treat methyl propanoate with LiAlH₄, identify reagent function, predict the product produced and the propose a step-wise mechanism of the reduction.

Example problem 10. Predict the reduction product of the following reaction and propose a mechanism for the reaction.

   Given: $CH_3CH_2CO_2CH_3$ and   1. $LiAlH_4$, 2. H_3O^+

   Find: Reduction product, reaction mechanism

   Plan: 1. Draw reactant structures
   2. Identify reactive center(s) in the substrate
   3. Identify reagent function based on structure

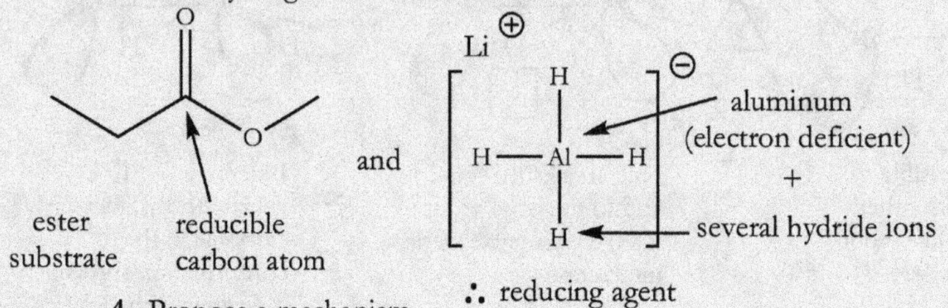

   ester   reducible
   substrate   carbon atom

   4. Propose a mechanism   ∴ reducing agent

   Solve:

nucleophilic acyl substitution

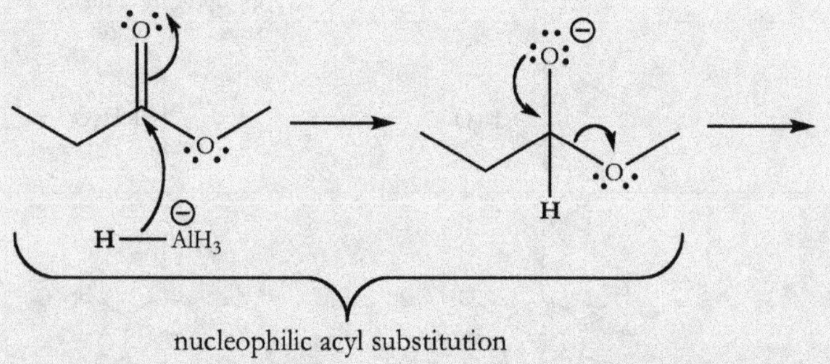

because the intermediate aldehyde product is more reactive than the starting ester, another equivalent of the hydride reduces the aldehyde to a primary alcohol:

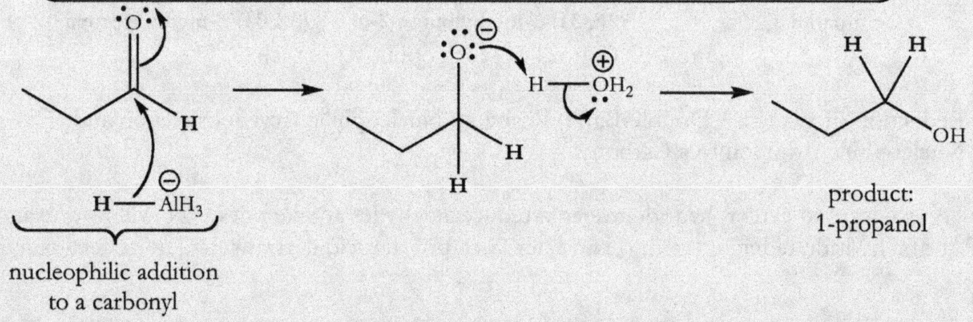

nucleophilic addition
to a carbonyl

product:
1-propanol

Check: 1. Reactant structures drawn correctly, 2. Reactive center identified in the substrate, 3. Reagent function identified based on structure, 4. Mechanism proposed using curved arrow formalism consistent with reduction of ester by hydride reagent.

Problem 7.13. Analyze (a) (i) and (ii) by electron pushing arrows and outline pathway. Then, predict the products and propose mechanisms for (b) (i) and (ii).

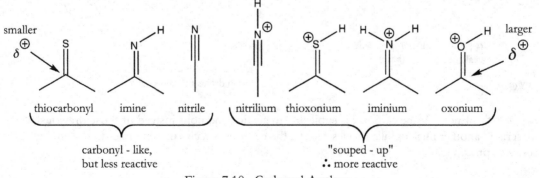

(a)

(i)

(ii)

(b)

(i)

draw intermediate

complete structure

(ii)

draw intermediate
I

draw intermediate
II

draw structure of
2-methyl-2-propanol

7.7.d. Carbonyl Analogs

A number of functional groups are found in organic chemistry that exhibit properties similar to carbonyls. These "disguised carbonyls" have reactivity patterns consistent with the carbonyl functionality and can undergo many of the same reactions. Some examples are shown in Figure 7.10.

smaller

larger

thiocarbonyl   imine   nitrile   nitrilium   thioxonium   iminium   oxonium

carbonyl - like,
but less reactive

"souped - up"
∴ more reactive

Figure 7.10. Carbonyl Analogs.

Problem 7.14. Analyze (a) by electron pushing arrows and outline pathway. Then, predict the product, draw the intermediate and propose a mechanism for (b).

(a)

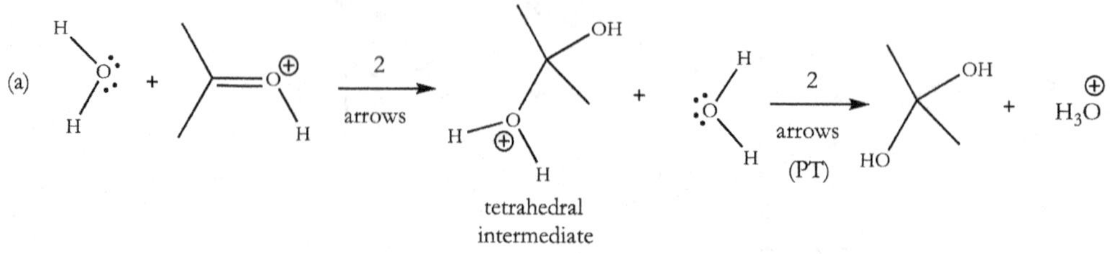

(b)

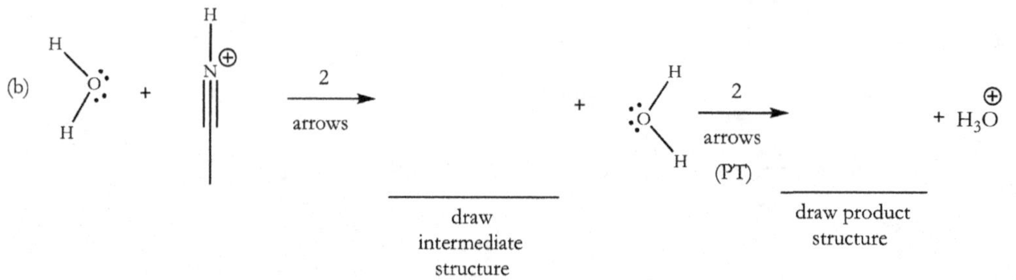

draw
intermediate
structure

draw product
structure

7.8 Carbonyl Nucleophiles: α-Substitution Reactions

The electron-withdrawing properties of the carbonyl group extend to the carbons adjacent or α to the C=O. These α carbons are more electropositive than normal alkyl groups. Furthermore, α carbon C-H bonds are weaker due to inductive electron withdrawal exerted by the C=O group, making them susceptible to abstraction by bases. Mechanisms for the formation of these carbonyl nucleophiles are outlined below:

❖ Enolates. In the presence of lithium diisopropylamide, a sufficiently strong, sterically hindered, non-nucleophilic base, deprotonation of the α protons (a typical acid-base reaction) can occur in the following fashion to yield a new nucleophile, the enolate:

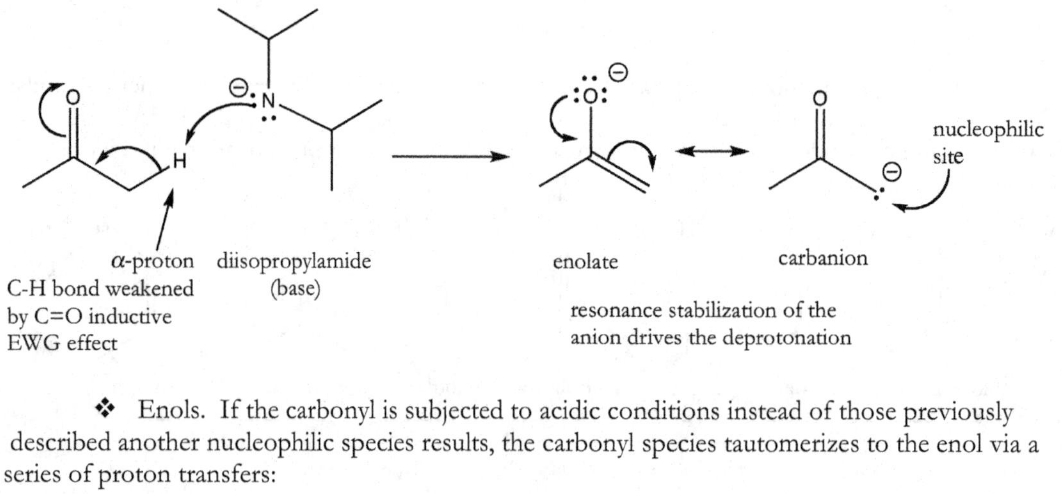

α-proton   diisopropylamide
C-H bond weakened   (base)
by C=O inductive
EWG effect

enolate

carbanion

nucleophilic
site

resonance stabilization of the
anion drives the deprotonation

❖ Enols. If the carbonyl is subjected to acidic conditions instead of those previously described another nucleophilic species results, the carbonyl species tautomerizes to the enol via a series of proton transfers:

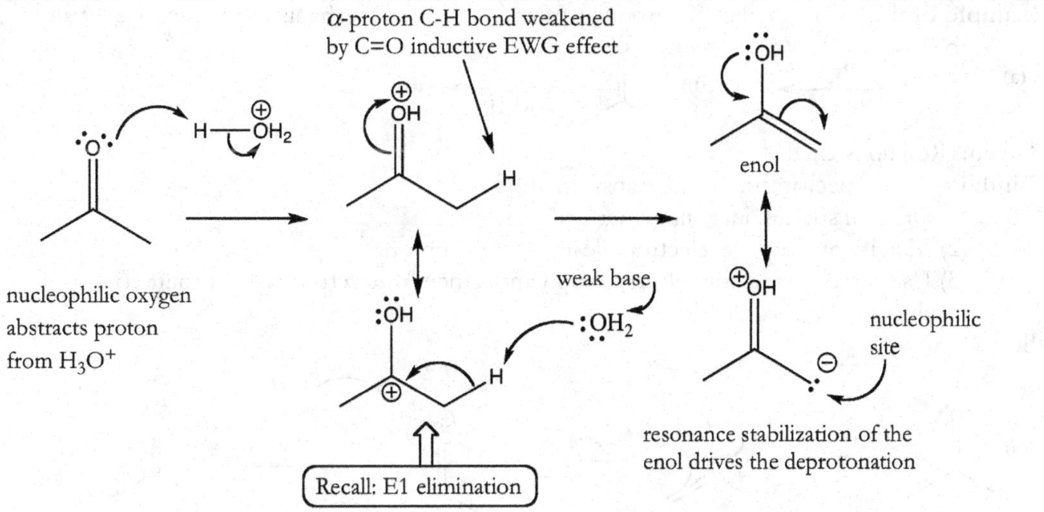

α-proton C-H bond weakened
by C=O inductive EWG effect

enol

nucleophilic oxygen
abstracts proton
from H_3O^+

weak base

nucleophilic
site

Recall: E1 elimination

resonance stabilization of the
enol drives the deprotonation

❖ Enamines. When a carbonyl undergoes nucleophilic addition as we saw in Section 7.7 by a 2° amine (R_2NH), the intermediate iminium ion (Figure 7.10) tautomerizes to an enamine via proton transfer.

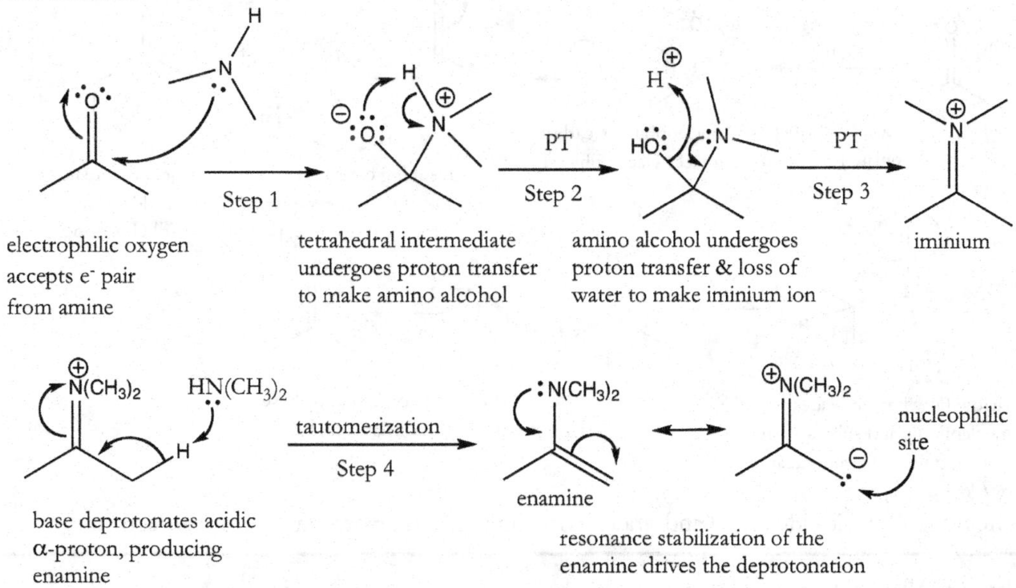

Step 1   Step 2 (PT)   Step 3 (PT)

electrophilic oxygen
accepts e⁻ pair
from amine

tetrahedral intermediate
undergoes proton transfer
to make amino alcohol

amino alcohol undergoes
proton transfer & loss of
water to make iminium ion

iminium

base deprotonates acidic
α-proton, producing
enamine

tautomerization
Step 4

enamine

nucleophilic
site

resonance stabilization of the
enamine drives the deprotonation

As you might note, the enol and enamine are neutral species and therefore, are less reactive than the negatively charged enolate. Regardless, enolates, enols and enamines are key intermediates in a host of reactions in organic chemistry: α – halogenation, α – alkylation and carbonyl condensation reactions to name a few. These processes share a common mechanistic theme: (1) generation of the α – carbonyl nucleophile and (2) nucleophilic attack on an electrophilic species. We will limit our focus to a few simple α-substitutions that best illustrate these commonalities.

7.8.a. α – Halogenation

Halogenation of enolates and enols may occur under basic or acidic conditions. Diatomic halogens – Cl_2, Br_2 and I_2, can be used to prepare α – monohalogenated ketones and aldehydes under acidic conditions. Under basic conditions, however, the reaction is difficult to stop after only one addition to methyl ketones; often the familiar Finkelstein reaction occurs wherein a trihalo carbonyl compound is formed that undergoes a subsequent nucleophilic acyl substitution. See Example Problem 11.

Example Problem 11. Predict the products of and propose a mechanism for reactions (a) and (b).

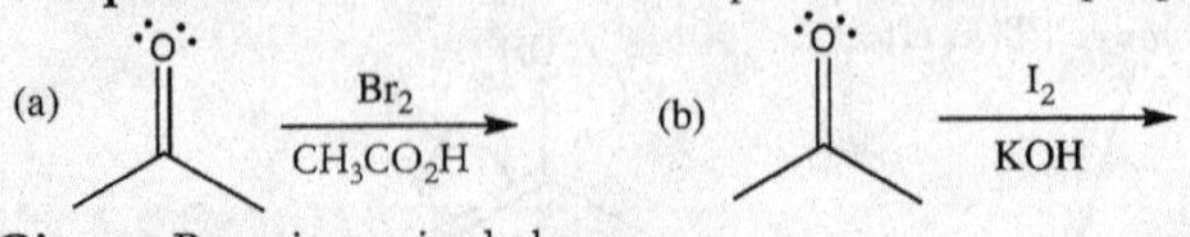

(a) $\xrightarrow[\text{CH}_3\text{CO}_2\text{H}]{\text{Br}_2}$ (b) $\xrightarrow[\text{KOH}]{\text{I}_2}$

Given: Reactions cited above.
Find: Products, mechanism for the transformation.
Plan: (1) Draw all species, including intermediates.
　　(2) Identify nucleophile, electrophile and leaving group.
　　(3) Use curved arrow formalism to show movement of electron pairs enroute to
　　product(s).
Solve:

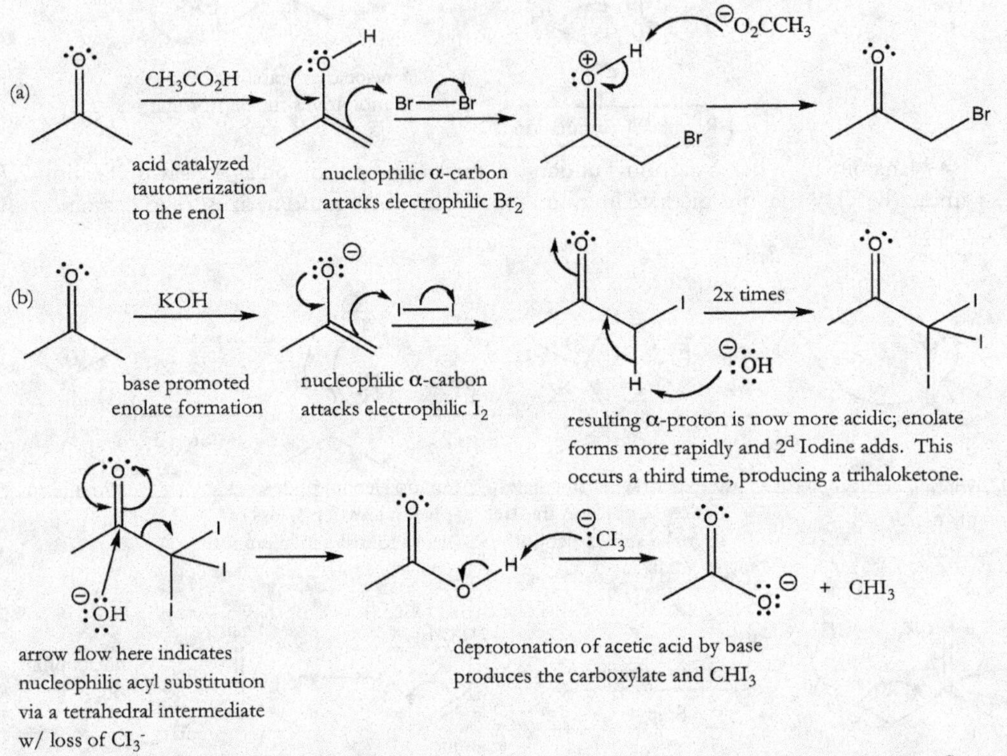

(a)

acid catalyzed
tautomerization
to the enol

nucleophilic α-carbon
attacks electrophilic Br₂

(b)

base promoted
enolate formation

nucleophilic α-carbon
attacks electrophilic I₂

resulting α-proton is now more acidic; enolate
forms more rapidly and 2^d Iodine adds. This
occurs a third time, producing a trihaloketone.

arrow flow here indicates
nucleophilic acyl substitution
via a tetrahedral intermediate
w/ loss of CI₃⁻

deprotonation of acetic acid by base
produces the carboxylate and CHI₃

Check: All species identified and drawn correctly; curved arrows show proper electron flow.

7.8.b. α – Alkylation

　　Carbonyl compounds are alkylated in a two-step process. The general reaction conducted
under basic conditions is shown below:

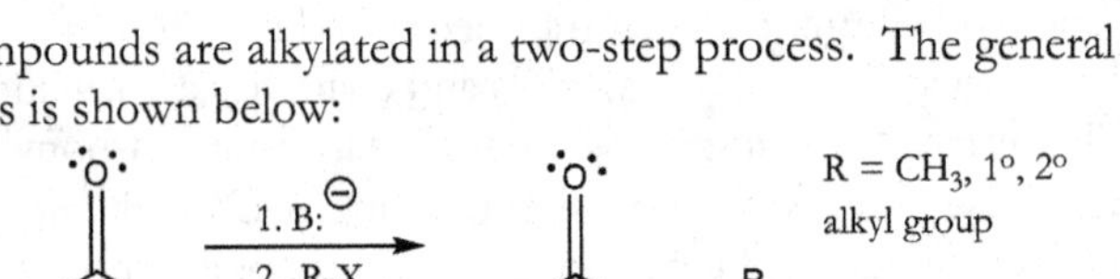

$\xrightarrow[\text{2. R-X}]{\text{1. B:}^{\ominus}}$

R = CH₃, 1°, 2°
alkyl group

X = Cl, Br, I, OTs

We have already examined the mechanistic sequence for this alkylation. Recall from earlier in this
section (1) enolate formation and (2) from Section 7.3, the S$_N$2 reaction between a Nuc: and R-X.
The four factors discussed relative to bimolecular nucleophilic substitution apply here – namely: the
nucleophile may be either neutral or negatively charged, suitable **substrates** are those which are not
sterically hindered: methyl > 1° > 2°, while containing a good **leaving group**, and polar, aprotic
solvents are generally preferred.

7.8.c. Carbonyl Condensations

When carbonyl nucleophiles react with other carbonyl species, a condensation occurs. Depending upon the carbonyl compound substrate, the products that result vary.

Substrate	Condensation	Product
Aldehyde	Aldol	β-hydroxy aldehyde
Ketone	Aldol	β-hydroxy ketone
Ester	Claisen	β-keto ester

In this section, we will examine the simplest cases of self-condensation between aldehydes and esters.

❖ General aldol self-condensation reactions, which may be conducted under either basic or acidic conditions, are shown below using two equivalents of acetaldehyde in our examples:

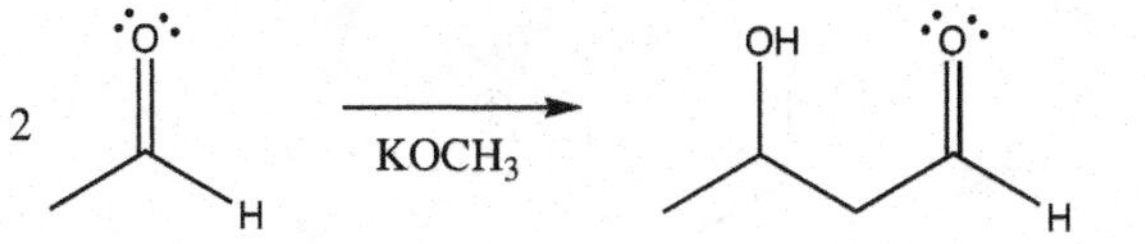

Important Points
- catalytic amount of base
- proceeds via enolate

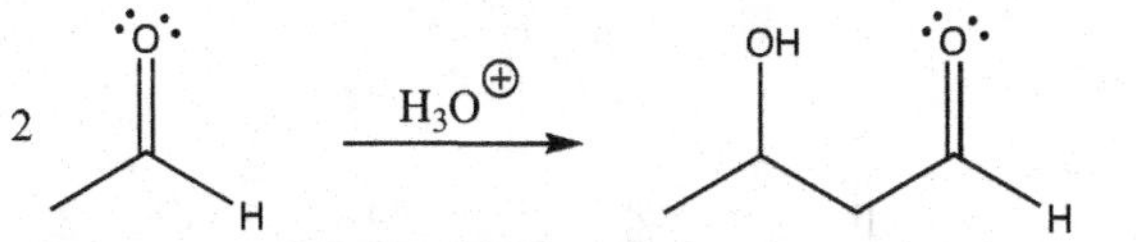

- catalytic amount of acid
- proceeds via enol

We have already examined the mechanistic sequence for this process. Recall from earlier in this section (1) enolate/enol formation and (2) from Section 7.7, the addition reaction between a Nuc: and C=O.

Of course, aldol reactions need not be self-condensations. Aldehydes and ketones may be used in mixed aldol reactions. Mixed or crossed aldol reactions do suffer from limitations however. If the carbonyl substrate(s) contain(s) more than one type of α-proton, complex mixtures of β-hydroxy products can occur. In addition, many examples of intramolecular aldol condensations are known. These reaction are used in synthesis to construct five- and six-membered ring systems that bear the ketone functionality.

Finally, the Knoevenagel condensation (aldol reaction coupled with β-hydroxy dehydration) provides α,β-unsaturated enones, species that are used in conjugate additions.

❖ The general Claisen self-condensation reaction, which is conducted under basic conditions, is shown below using methyl acetate as our example:

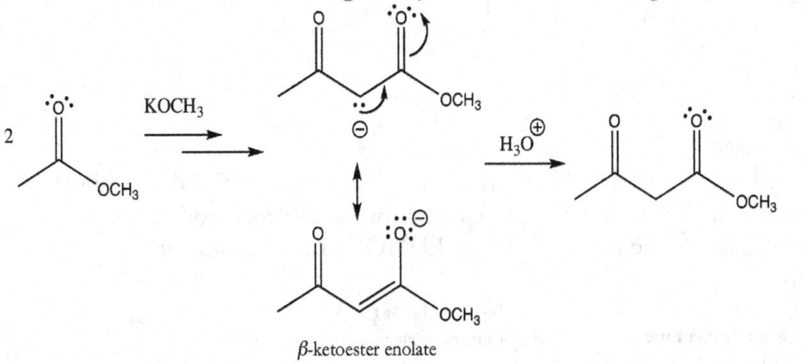

β-ketoester enolate

Important Points
- full equivalent of base
- base used is normally the same as the leaving group to minimize trans-esterification
- proceeds via enolate
- acidification of β-ketoester enolate required to obtain product

We have already examined the mechanistic sequence for this process. Recall from earlier in this section (1) enolate/enol formation, (2) from Section 7.7, the acyl substitution reaction between a Nuc: and C=O bearing a leaving group (-OCH₃ in our example) and (3) proton transfer to the β-ketoester enolate yielding the β-ketoester product.

Like the aldol reaction, Claisen reactions need not be self-condensations. Esters and ketones may be used in mixed Claisen reactions. They can suffer from the same drawbacks as the mixed aldol reactions. If the carbonyl substrate(s) contain(s) more than one type of α-proton, complex mixtures of β-keto ester products can occur. In addition, examples of intramolecular Claisen condensations (the Dieckmann condensation) are known. This reaction is used to construct cyclic β-keto esters.

Problem 7.15. Predict the products of the following reactions.

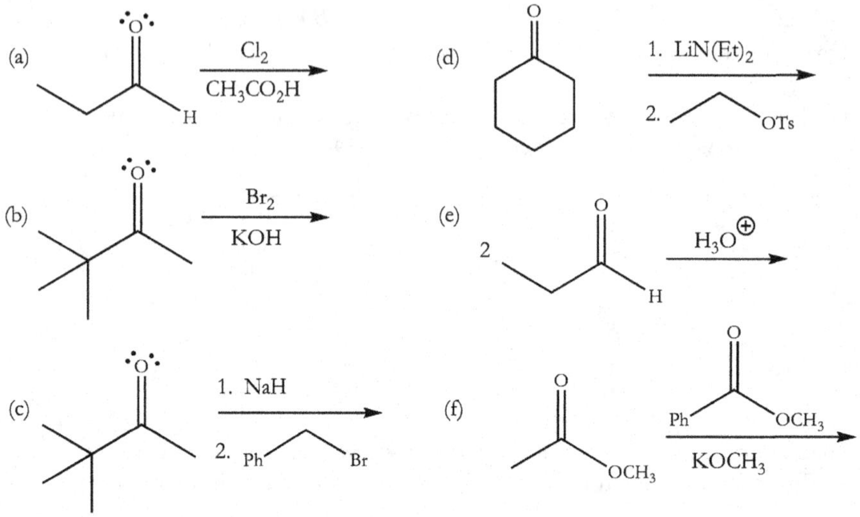

(a)
$$\frac{Cl_2}{CH_3CO_2H}$$

(d)
1. LiN(Et)$_2$
2. ⌁OTs

(b)
$$\frac{Br_2}{KOH}$$

(e)
2 ⌁$\xrightarrow{H_3O^{\oplus}}$

(c)
1. NaH
2. Ph⌁Br

(f)
Ph⌁OCH$_3$ $\xrightarrow{KOCH_3}$

7.9 The Stork Reaction: Enamine Addition to α,β-Unsaturated Carbonyls

Our final installment in the carbonyl chemistry section is an extension of the insight we have just gained concerning the reactions of carbon nucleophiles with electrophilic species coupled with our knowledge of proton transfers and carbonyl analogs. The overall process is outlined below:

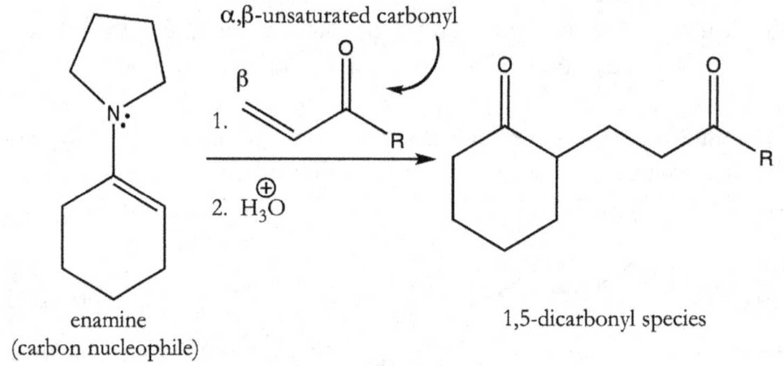

α,β-unsaturated carbonyl

1.
2. H$_3$O$^{\oplus}$

enamine
(carbon nucleophile)

1,5-dicarbonyl species

Why in this example, does the carbon nucleophile selectively attack the β-carbon of the carbonyl rather than C=O? The answer is found by examining the nature of the molecular orbitals of the reacting species – the enamine nucleophile HOMO and the LUMO of the electrophilic α,β-unsaturated carbonyl species.

large lobe of enamine
HOMO is on carbon 2

bigger lobe at C4 than C=O;
better overlap w/ Nuc at C4

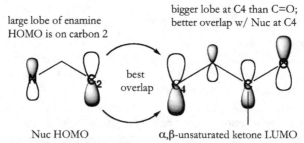

best
overlap

Nuc HOMO

α,β-unsaturated ketone LUMO

Bond formation is a question of where the best overlap occurs. In this case, the HOMO-LUMO interaction most likely to lead to a bonding interaction is overlap that occurs between the lobe at C_2 of the enamine HOMO and the lobe at C_4 of the α,β-unsaturated carbonyl LUMO.

❖ The Mechanism of the Stork Reaction

This reaction employs several of the mechanistic sequences we have examined in this chapter. Here is the breakdown and step-by-step pathway below:

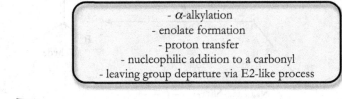

- α-alkylation
- enolate formation
- proton transfer
- nucleophilic addition to a carbonyl
- leaving group departure via E2-like process

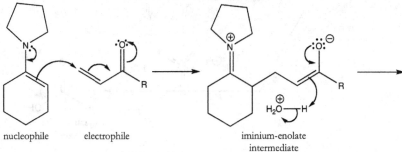

nucleophile   electrophile   iminium-enolate intermediate

Step 1: α-alkylation   **Step 2:** proton transfer quenches enolate

Step 3: nucleophilic addition to a carbonyl analog   tetrahedral intermediate

Step 4: proton transfer to amine

tetrahedral intermediate   1,5-dicarbonyl species

Step 5: proton transfer to solvent collapses tetrahedral intermediate, releasing amine leaving group and forming the 1,5-dicarbonyl product

7.10. Oxidation of Alcohols with Cr[VI] Reagents

Many chemical transformations may be classified as oxidation reactions. As you learned from general chemistry, a chemical species undergoes oxidation if it experiences a net loss of electrons in a reaction.

Organic chemistry applies this definition in a slightly different manner. Oxidation is defined as the net gain of oxygen atoms or loss of hydrogen atoms from an organic molecule. See Appendix B for a list of common oxidizing agents. Primary (1°) and secondary (2°) alcohols undergo oxidation to carbonyl compounds when treated with chromium reagents. The general reactions are shown in Figure 7.11:

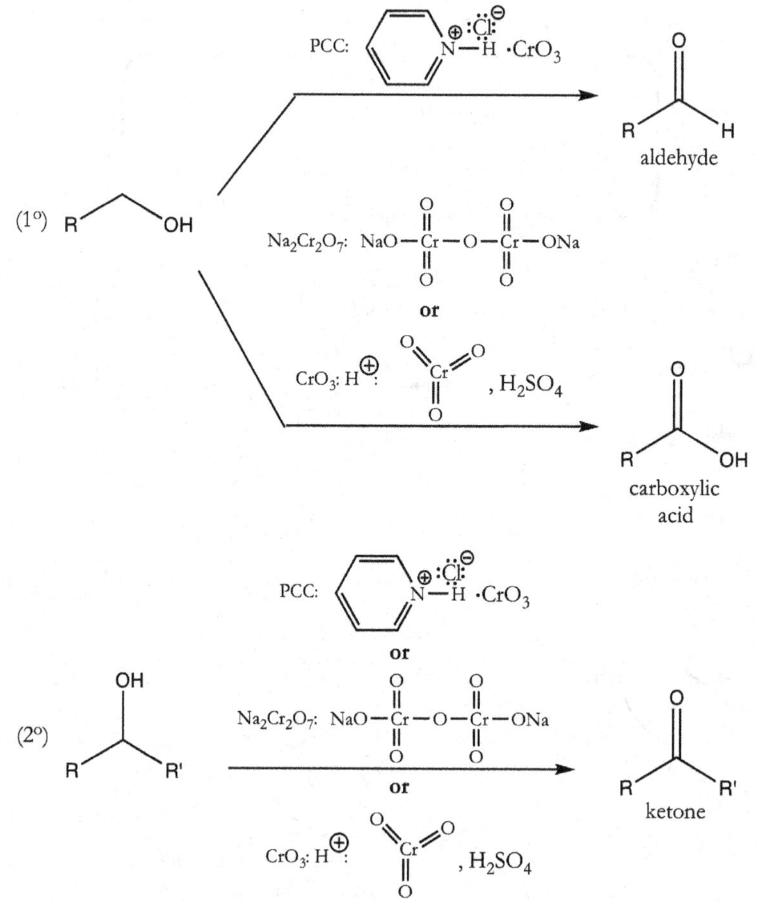

Figure 7.11. Oxidation of Alcohols by Cr(VI) Reagents.

➤ Partial oxidation of 1° alcohols: Selectivity of pyridinium chlorochromate (PCC)

Pyridinium chlorochromate (PCC) was developed to ameliorate the tendency of Cr(VI) reagents in strongly acidic media to oxidize 1° alcohols to carboxylic acids. The weakly acidic pyridinium ion, soluble in CH_2Cl_2, serves as a proton source to protonate the chromate ester. Its conjugate base, pyridine, then abstracts the β-proton to eliminate the chromium species in the oxidation step. The absence of water in the reaction medium limits the oxidation of primary alcohols to an aldehyde. The step-by-step pathway for this process is characterized by processes we have examined earlier:

> - nucleophilic attack on an electrophile producing a chromate ester
> - proton transfer
> - departure of chromium species leaving group and oxidation via an E2-like process

Here is a mechanism for the oxidation of 1-butanol to 1-butanal by PCC.

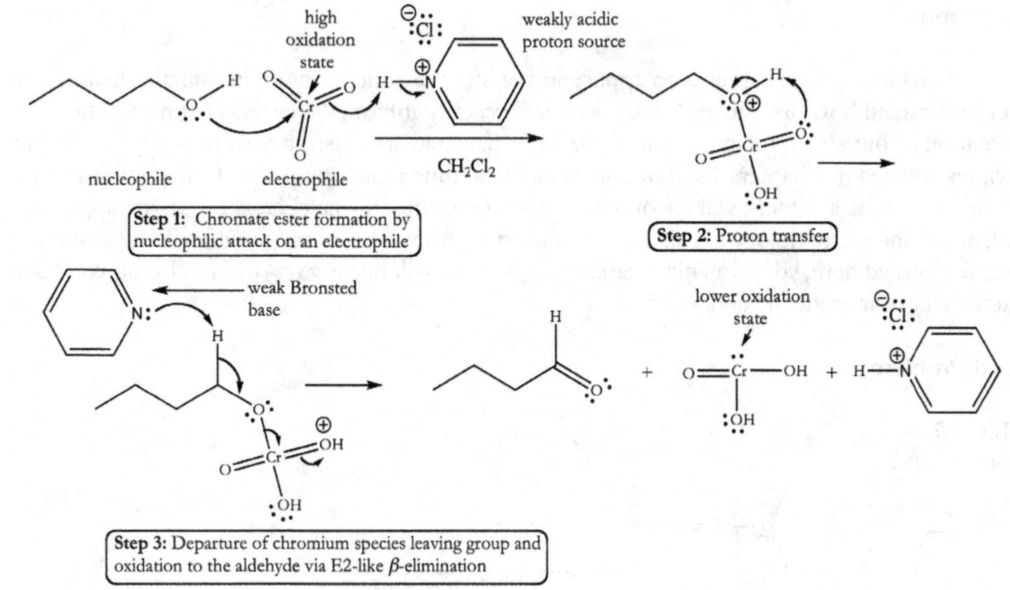

high oxidation state

weakly acidic proton source

nucleophile electrophile

CH₂Cl₂

Step 1: Chromate ester formation by nucleophilic attack on an electrophile

Step 2: Proton transfer

weak Brønsted base

Step 3: Departure of chromium species leaving group and oxidation to the aldehyde via E2-like β-elimination

lower oxidation state

➤ Oxidation of 1° alcohols: Nonselectivity of Cr(VI) reagents in acid media

As noted in Figure 7.11, the powerful Cr(VI) oxidizing agents such as CrO_3, H_2SO_4 and $Na_2Cr_2O_7$ oxidize primary alcohols to carboxylic acids. The oxidation process, which involves several reactions we have discussed in this chapter is outlined below:

> - chromate ester formation by nucleophilic attack on an electrophile (2)
> - proton transfer (2)
> - nucleophilic addition to a carbonyl
> - leaving group departure via E2-like process (2)

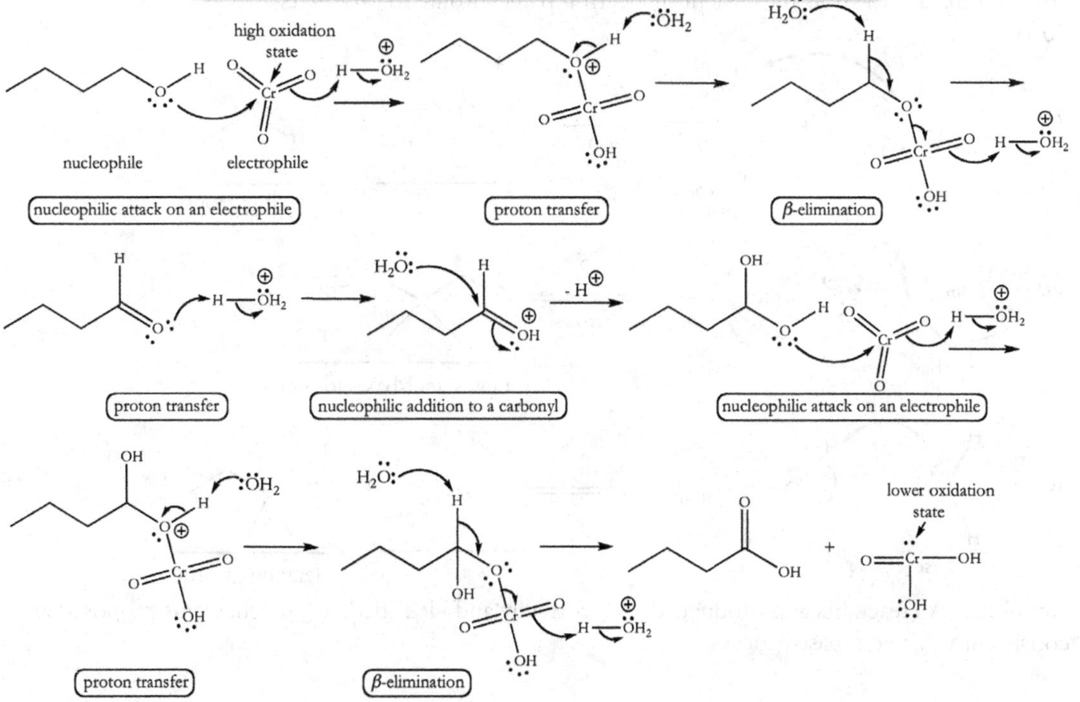

high oxidation state

nucleophile electrophile

nucleophilic attack on an electrophile

proton transfer

β-elimination

proton transfer

nucleophilic addition to a carbonyl

nucleophilic attack on an electrophile

proton transfer

β-elimination

lower oxidation state

7.11. Summary

At its heart, this chapter is an application of structure and function in organic chemistry to better understand how organic molecules, ions and reactive intermediates react. It marks the culmination of our study; the mechanisms we have dissected are illustrative of not only key chemical principles covered in Chapters 1-6, but also of essential functional group transformations and types of reactions you, as a student will encounter in undergraduate organic chemistry. If you apply the problem solving methodology consistently as shown in this work and use your knowledge of the principles learned here, your organic chemistry experience will be very rewarding. I wish you good fortune in all your academic endeavors.

Solved Problems

Problem 7.1.
Given:

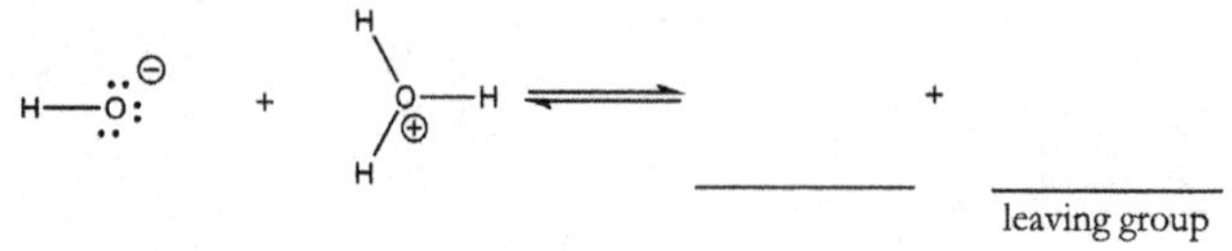

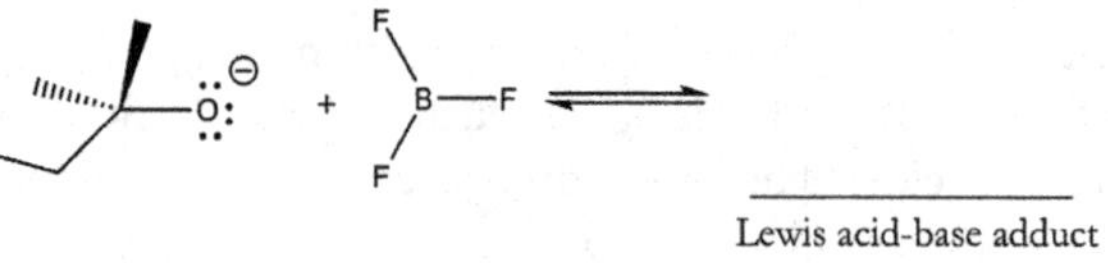

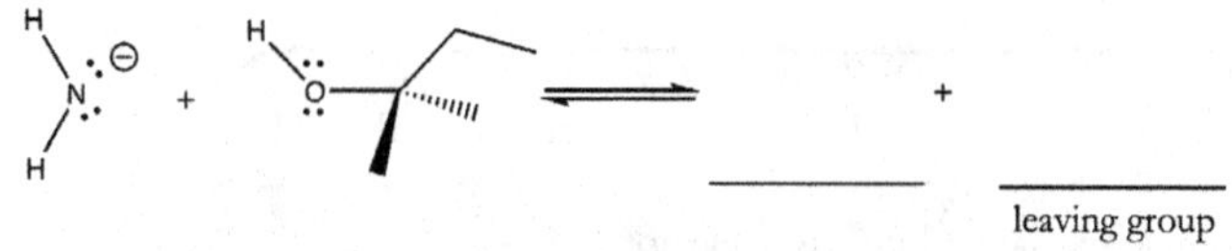

Find: (1) mechanism for the transformation, (2) reaction products.
Plan: (1) Draw all species. (2) Identify base, acid and leaving group for the reaction. (3) Use curved arrow formalism to show movement of electron pairs enroute to product(s).
Solve:

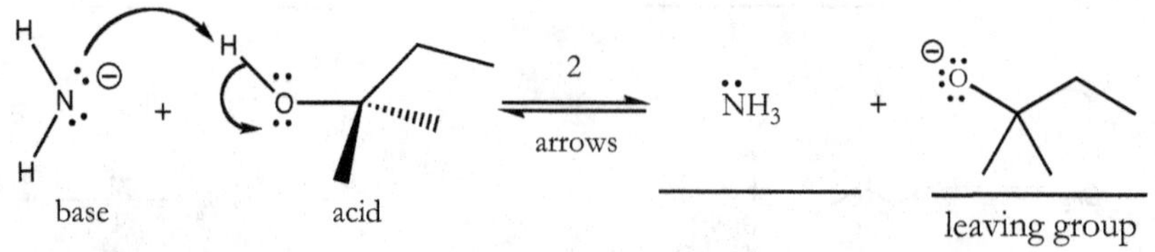

Check: (1) All reactants and products drawn correctly and identified, (2) mechanisms proposed are consistent w/an acid-base process.

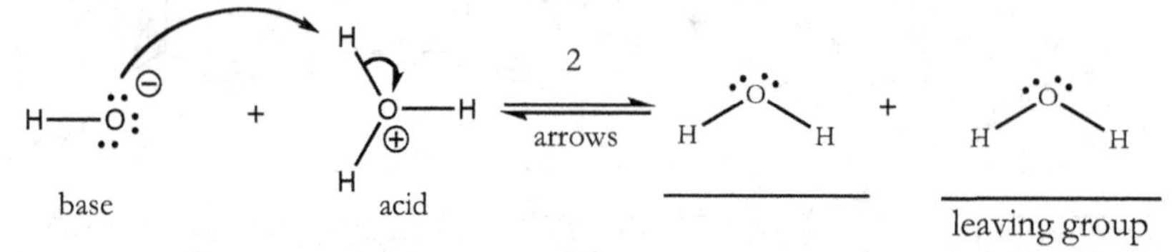

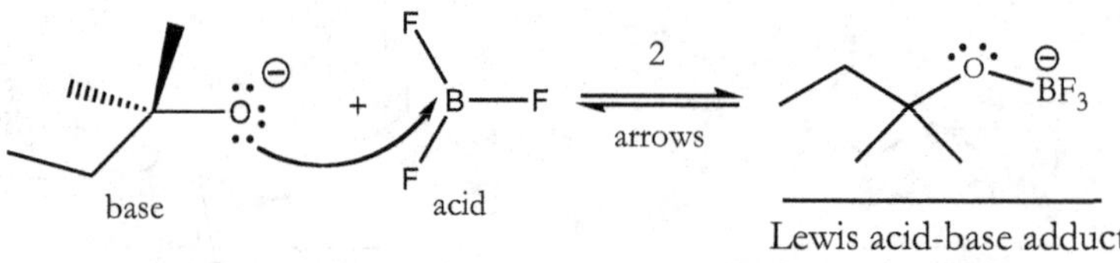

110

Problem 7.2.
Given:

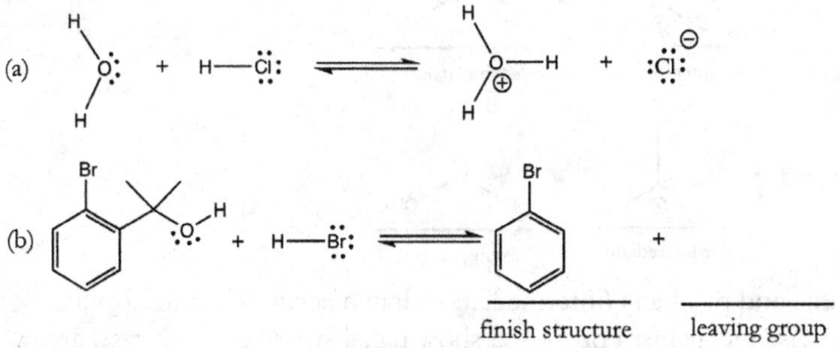

(a)

(b)

finish structure   leaving group

Find: (1) mechanism for the transformations in (a) and (b), (2) reaction products for (b).
Plan: (1) Draw all species. (2) Identify base, acid and leaving group for the reaction.
(3) Use curved arrow formalism to show movement of electron pairs enroute to product(s).
Solve:

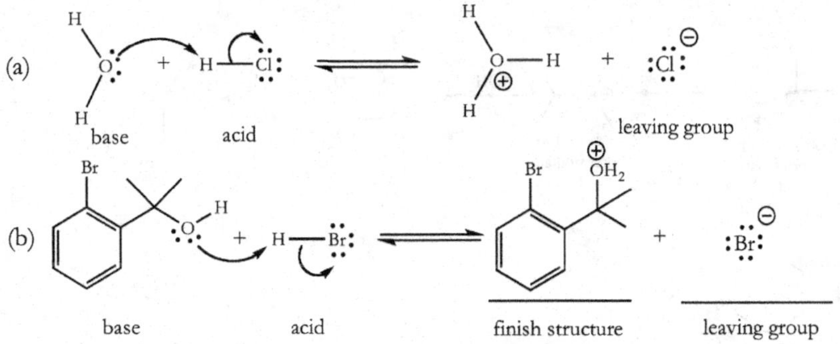

(a)

base   acid   leaving group

(b)

base   acid   finish structure   leaving group

Check: (1) All reactants and products drawn correctly and identified, (2) mechanisms proposed are consistent w/an acid-base process.

Problem 7.3.
Given:

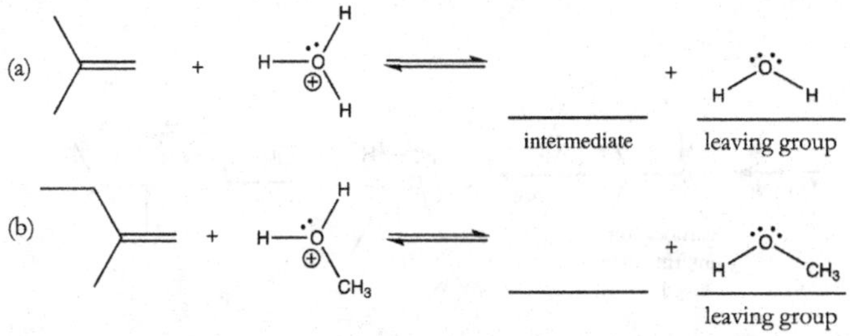

(a)

intermediate   leaving group

(b)

leaving group

Find: Electron-pushing arrows (electron flow) required to produce the intermediates.
Plan: (1) Draw structures for the reactants and products; identify the nucleophile and electrophile.
(2) Write the stepwise mechanism showing all reactants, intermediates (products) and leaving groups using curved arrow formalism to show e⁻ pair movements.
Solve:

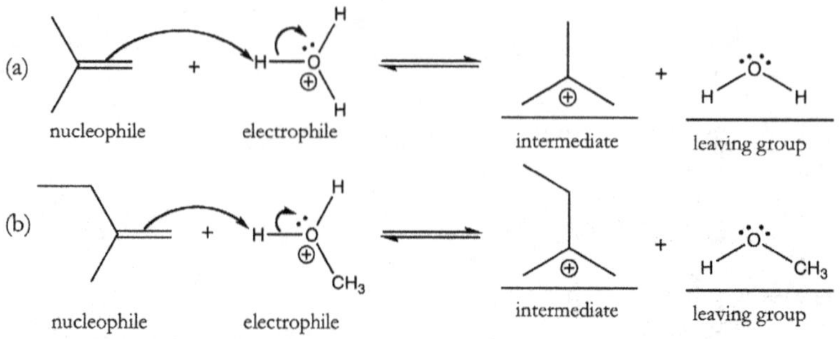

(a) nucleophile — electrophile — intermediate — leaving group

(b) nucleophile — electrophile — intermediate — leaving group

Check: (1) Structures for all reactants and products (intermediates) drawn accurately; nucleophile and electrophile identified. (2) Stepwise mechanism proposed showing all species. (3) Curved arrow formalism shows e⁻ pair movements leading to the most stable carbocation product.

Problem 7.4.
Given:

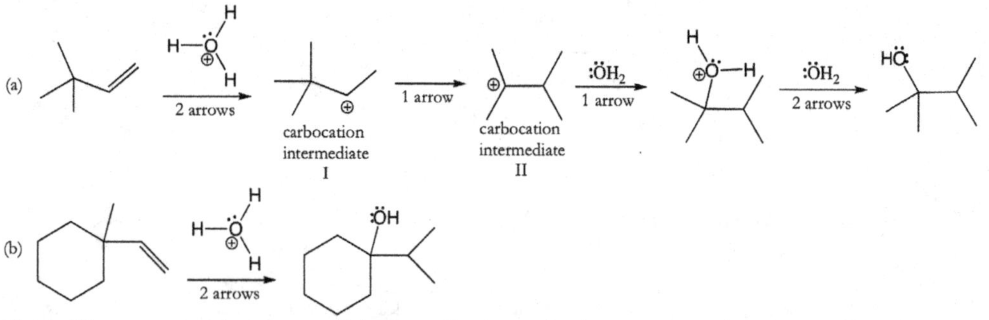

(a) 2 arrows — carbocation intermediate I — 1 arrow — carbocation intermediate II — 1 arrow — 2 arrows

(b) 2 arrows

Find: Electron-pushing arrows (electron flow) required to produce the intermediates and products.
Plan: (1) Identify the nucleophile and electrophile in reactions (a) and (b). (2) Write the stepwise mechanism for (a) using curved arrow formalism to show e⁻ pair movements. (3) Write the stepwise mechanism for (a) showing all reactants, intermediates and products using curved arrow formalism to show e⁻ pair movements.
Solve:

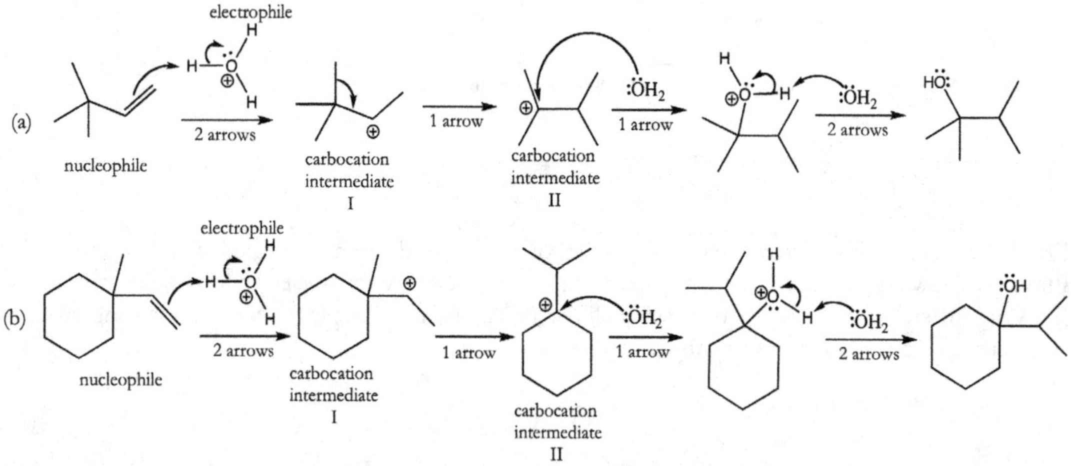

(a) nucleophile — 2 arrows — carbocation intermediate I — 1 arrow — carbocation intermediate II — 1 arrow — 2 arrows

(b) nucleophile — 2 arrows — carbocation intermediate I — 1 arrow — carbocation intermediate II — 1 arrow — 2 arrows

Check: (1) Structures for all reactants and products (intermediates) drawn accurately; nucleophile and electrophile identified. (2) Stepwise mechanism proposed showing all species. (3) Curved arrow formalism shows e⁻ pair movements including carbocation rearrangement leading to the most stable carbocation product.

112

Problem 7.5.
Given:

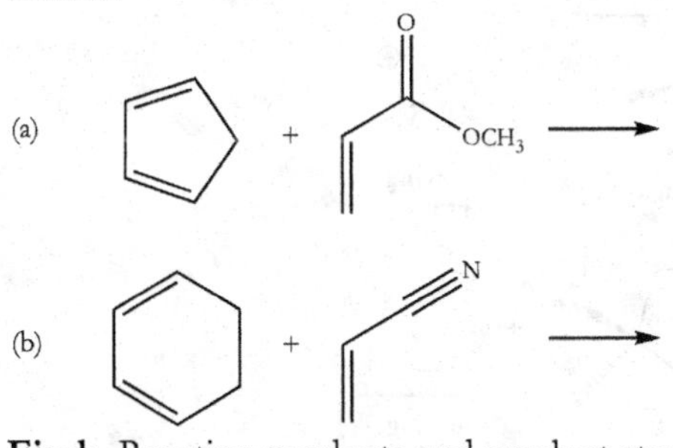

(a)

(b)

Find: Reaction products and product stereochemistry.
Plan: (1) Identify diene and dienophile for the reactions. (2) Use the your knowledge of the reaction mechanism to determine the product structure and stereochemistry. (3) Complete/draw all species.
Solve:

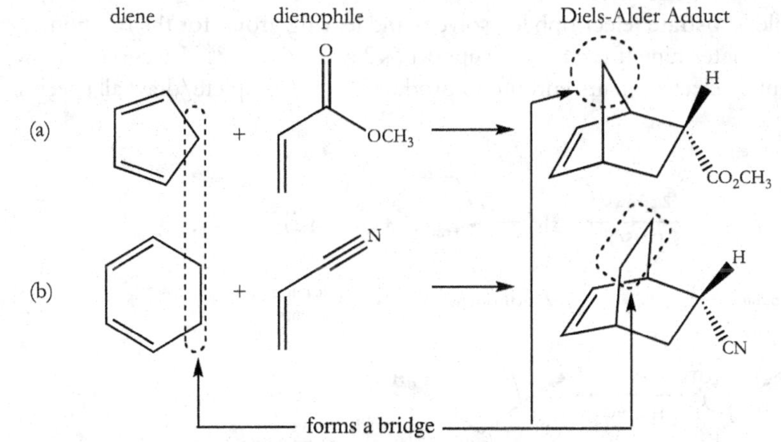

diene   dienophile   Diels-Alder Adduct

(a)

(b)

forms a bridge

In these cases, when the cyclic 1,3-dienes react with the dienophile, the circled diene carbons form a bridge in the product. The dienophile orients itself during the reaction to have the large groups as far away from the bridge as possible. Therefore, the Diels-Alder adducts have the stereochemistry indicated.

Check: (1) All species drawn correctly and identified, (2) Diels-Alder adducts (products) consistent w/a mechanism that involves a cyclic transition state. (3) Stereochemistry in product correctly displayed.

Problem 7.6.
Given:

(a) :N≡C:⁻ + [S-2-tosylbutane with H and OTs] → NC—[R-2-methylbutanenitrile] + TsÖ:⁻
2 arrows / DMSO

S-2-tosylbutane   *R*-2-methylbutanenitrile   leaving group

(b) :N̈=N⁺=N̈:⁻ + [cyclopentane ring with OTs and H] → [product] +
HMPA

(*R*) stereochemistry   ( ) stereochemistry   leaving group

Complete structure

Find: (1) mechanism for the transformation in (a), (2) mechanism for the transformation in (b), reaction products and product stereochemistry
Plan: (1) Identify nucleophile, substrate (electrophile), solvent and leaving group for the reactions. (2) Use the S$_N$2 factor guide to determine if the factors support S$_N$2 processes. (3) Use curved arrow formalism to show movement of electron pairs enroute to product(s). (4) Complete/draw all species.
Solve:

nucleophile   electrophile

(a) :N≡C:⁻ + [H, ÖTs substrate] → NC—[product] + TsÖ:⁻
2 arrows / DMSO

S-2-tosylbutane   *R*-2-methylbutanenitrile   leaving group

nucleophile   electrophile

(b) :N̈=N⁺=N̈:⁻ + [ring with ÖTs] → [ring with N₃] + TsÖ:⁻
HMPA

(*R*) stereochemistry   (*S*) stereochemistry   leaving group

Complete structure

Factors supporting S$_N$2: (1) Nucleophiles are weak bases and good e⁻ pair donors, (2) substrates (electrophiles) are 2° alkyl tosylates, (3) solvents are polar, aprotic, (4) tosylates are weak bases; ∴ good leaving groups.
Check: (1) All species drawn correctly and identified, (2) All factors support S$_N$2 process, (3) bimolecular mechanism proposed consistent w/an S$_N$2 process. (4) Inversion of stereochemistry in product observed.

Problem 7.7.
Given:

(a) [structure with —OH₂⁺] —1→ [structure with ⁺] + H₂O   :Ï:⁻ → [structure with —I]

(b) [structure with —S(CH₃)₂⁺] —1→ [ ] + [ ]   :CN⁻ → [structure with —CN]

Find: (1) mechanism for the transformation in (a) and (b), (2) reaction intermediates.
Plan: (1) Draw all species, including intermediates. (2) Identify nucleophile, substrate (electrophile), leaving group and solvent for the reaction. (3) Use the S$_N$1 factor guide to determine how the four

114

factors will support an S_N1 process. (4) Use curved arrow formalism to show movement of electron pairs enroute to product(s).

Solve:

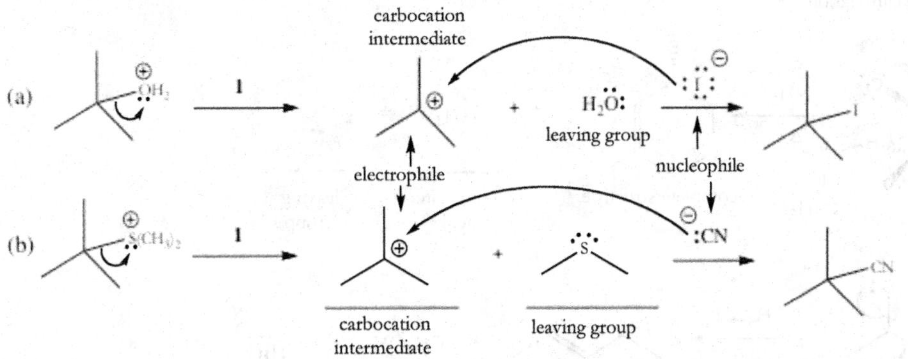

Factors supporting S_N1: (1) substrates – (a) is a protonated 3° alcohol; (b) is a dimethylated 3° sulfide, (2) assume solvents are polar, protic, (3) leaving groups are neutral molecules, ∴ good leaving groups. (4) Nucleophiles are weak bases (good e⁻ pair donors) and ∴ E1 would not be expected.

Check: (1) All reactants, intermediates and products drawn correctly and identified, (2) All factors support S_N1 process, (3) unimolecular mechanism proposed consistent w/ S_N1 process.

Problem 7.8.

Given:

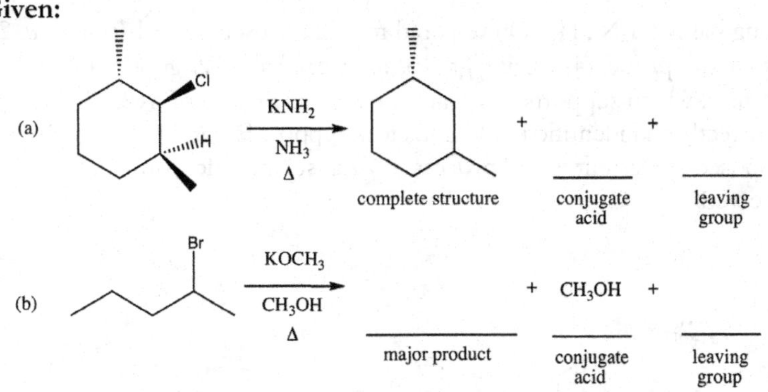

Find: (1) mechanism and reaction products for (a) and (b).

Plan: (1) Draw all species. (2) Identify base, substrate β-protons, leaving group and solvent for the reaction. (3) Use the E2 factor guide to determine how the four factors will support an E2 process. (4) Use curved arrow formalism to show movement of electron pairs enroute to product.

Solve:

β-proton is not anti periplanar to leaving group, ∴ does not participate in the elimination

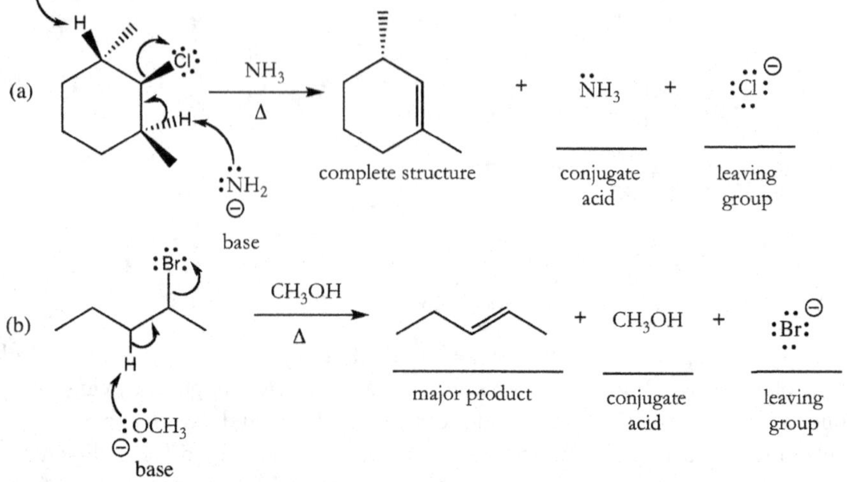

(a)

NH_3
Δ

complete structure

+ $\ddot{N}H_3$ +

conjugate acid

$:\ddot{Cl}:^{\ominus}$

leaving group

base

(b)

CH_3OH
Δ

major product

+ CH_3OH +

conjugate acid

$:\ddot{Br}:^{\ominus}$

leaving group

base

β-proton on the most highly substituted carbon is anti periplanar to leaving group. Its abstraction leads to the Zaitsev product, the major product observed.

Factors supporting E2: (1) Strong bases (H_2N^-, H_3CO^-) required for E2 are used, (2) substrates are 2° alkyl halides, (3) solvents are polar and protic, (4) leaving groups are weak bases, ∴ good leaving groups, (5) high temperature is used which supports elimination reactions instead of S_N2.

Check: (1) All species drawn correctly and identified, (2) All factors support E2 process, (3) bimolecular mechanism proposed consistent w/E2 process. (4) Zaitsev product formed; trans stereochemistry in product observed.

Problem 7.9.
Given:

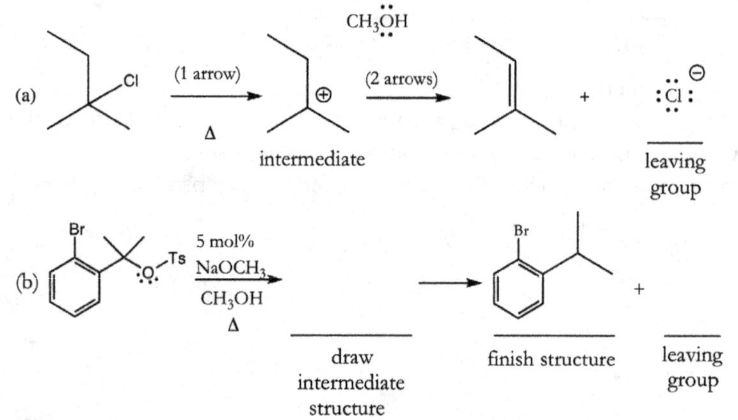

$CH_3\ddot{O}H$

(a)

(1 arrow)
Δ

intermediate

(2 arrows)

+ $:\ddot{Cl}:^{\ominus}$

leaving group

(b)

5 mol%
$NaOCH_3$
CH_3OH
Δ

draw intermediate structure

finish structure

+

leaving group

Find: (1) mechanism for (a), (2) structures of all species and mechanism for (b).
Plan: (1) Draw all species, including intermediates. (2) Identify base, substrate β-protons, leaving group and solvent for the reaction. (3) Use the E1 factor guide to determine how the four factors will support an E1 process. (4) Use curved arrow formalism to show movement of electron pairs enroute to product.

116

Solve:

β-proton   CH₃ÖH ← base & solvent

(a)   Cl: (1 arrow) ⟶ Δ   intermediate (2 arrows) ⟶   +   :Cl:⊖
                                                              leaving group

(b)   [structure with Br, Ts, O] CH₃OH / solvent, Δ ⟶ draw intermediate structure ⟶ finish structure   +   :ÖTs⊖
      β-proton :ÖCH₃⊖ base   leaving group

Factors supporting E1: (1) Weak base (CH₃OH) used in (a) and small mole ratio of base (H₃CO⁻) used in (b), (2) substrate in (a) is a 3° alkyl halide, while in (b) an alkyl tosylate, (3) solvents are polar and protic, (4) leaving groups are weak bases, ∴ good leaving groups, (5) high temperature is used which supports elimination reactions instead of S$_N$1.

Check: (1) All species and intermediate drawn correctly and identified, (2) All factors support E1 process, (3) unimolecular mechanism proposed consistent w/ E1 process.

Problem 7.10.

Given: Friedel-Krafts acylation of benzene via formation of the acylium ion, R-C≡O⁺.

Find: (1) mechanism for the formation of the acylium ion, R-C≡O⁺, (2) draw a resonance structure that has the positive formal charge on carbon.

Plan: (1) Draw all species, including intermediates. (2) Identify the Lewis acid and base, and Lewis acid-base adduct for the reaction. (3) Use the electrophilic activation step of the Friedel-Krafts alkylation EAS sequence as the starting point for the mechanism. (4) Use curved arrow formalism to show movement of electron pairs enroute to the acylium ion. (5) Use resonance theory to draw resonance contributors.

Solve:

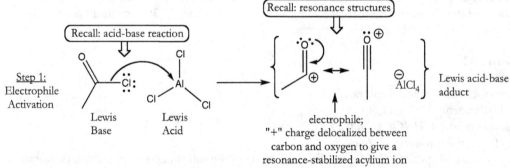

Step 1: Electrophile Activation

Lewis Base   Lewis Acid

electrophile; "+" charge delocalized between carbon and oxygen to give a resonance-stabilized acylium ion

Lewis acid-base adduct

Check: (1) All species and intermediate drawn correctly and identified, (2) Step 1 (electrophile activation) of the Friedel-Krafts alkylation EAS used as a template for the mechanism, (3) mechanism proposed is consistent w/Lewis acid-base theory.

Problem 7.11.

Given: Sulfonation of benzene via formation of a protonated sulfur trioxide ion, $^+SO_3H$.

Find: (1) Electrophilic activation step leading to the protonated sulfur trioxide ion, $^+SO_3H$, (2) mechanistic sequence for the sulfonation of benzene.

Plan: (1) Draw all species, including intermediates. (2) Identify the Lewis acid and base, and Lewis acid-base adduct for the reaction. (3) Use the electrophilic activation step of benzene nitration EAS sequence as the starting point for the mechanism. (4) Use curved arrow formalism to show movement of electron pairs enroute to the acylium ion. (5) Use resonance theory to draw resonance contributors.

Solve:

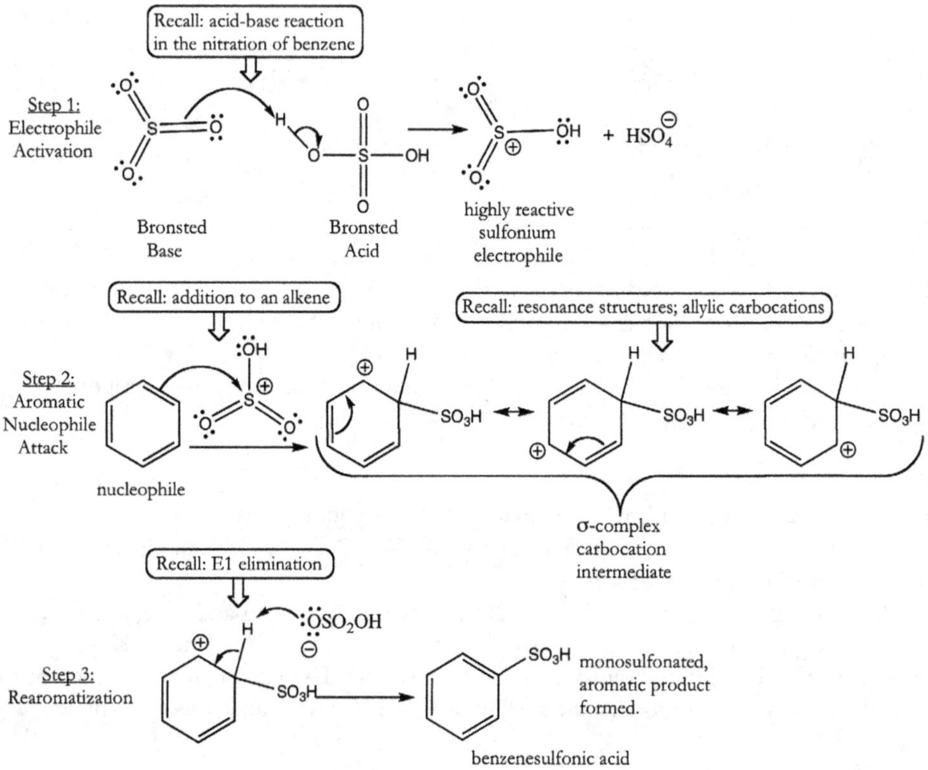

Check: (1) All species and intermediate drawn correctly and identified, (2) Step 1 (electrophile activation) of the nitration EAS used as a template for the mechanism, (3) mechanism proposed is consistent w/Lewis acid-base and EAS theory.

Problem 7.12.

Given:

(a) bromobenzene + acetyl chloride/$AlCl_3$ →

(b) methylbenzene + SO_3/H_2SO_4 →

(c) chlorobenzene + HNO_3/H_2SO_4 →

Find: (1) reaction products for (a-c).

Plan: (1) Draw all species. (2) Identify directing ability of the substituents on the aromatic reactants. (3) Based on knowledge of EAS, predict the product(s).

118

Solve:

(a)

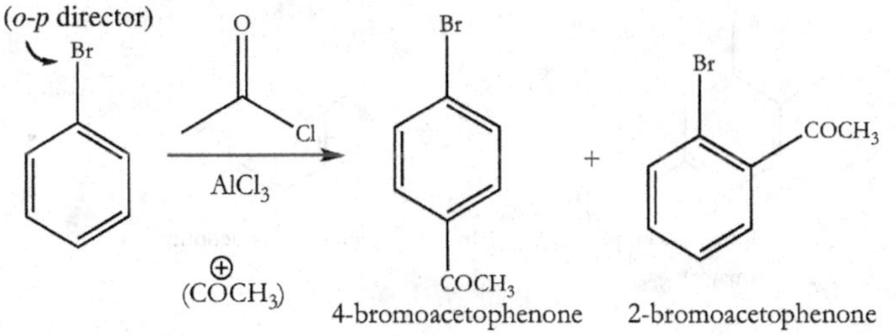

(*o-p* director)

4-bromoacetophenone   2-bromoacetophenone

(b)

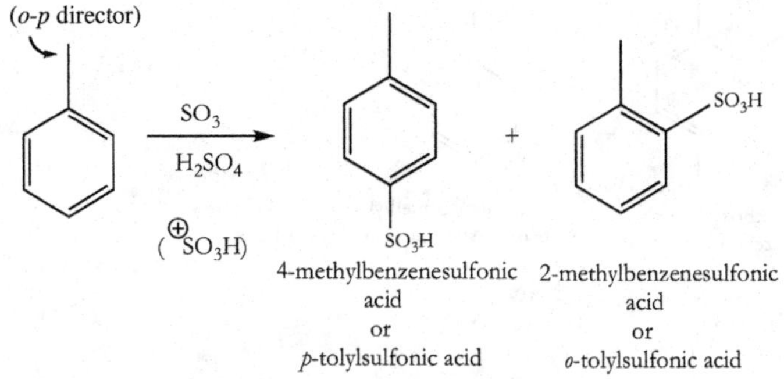

(*o-p* director)

4-methylbenzenesulfonic   2-methylbenzenesulfonic
acid   acid
or   or
p-tolylsulfonic acid   *o*-tolylsulfonic acid

(c)

(*o-p* director)

4-chloro-1-nitrobenzene   2-chloro-1-nitrobenzene
or   or
p-chloronitrobenzene   *o*-chloronitrobenzene

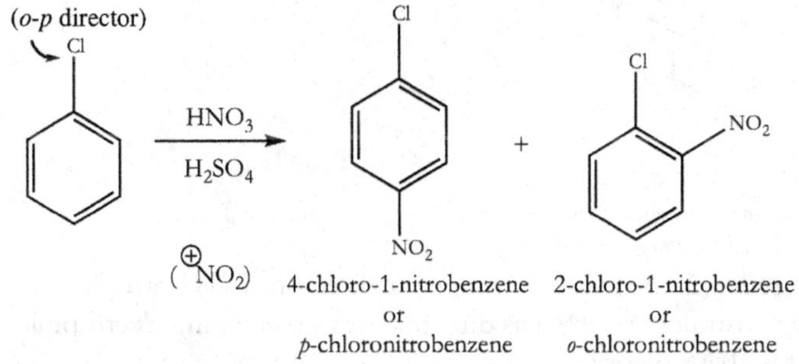

Check: (1) Structures drawn correctly, (1) *o-p* directors and electrophilic species identified, (3) products proposed are consistent w/EAS theory.

Problem 7.13.
Given:
(a) 1-bromo-2-methylbenzene (2-bromotoluene) + acetyl chloride/AlCl₃ →
(b) 1-methoxy-3-nitrobenzene (3-nitroanisole) + SO₃/H₂SO₄ →
(c) 4-chlorobenzene sulfonic acid + HNO₃/H₂SO₄ →
Find: Reaction products for (a-c).
Plan: (1) Draw all species. (2) Identify directing ability of the substituents on the aromatic reactants. (3) Select the most activating substituent as director for the incoming electrophile. (4) Based on knowledge of EAS, predict the product(s).

119

Solve:

(a)

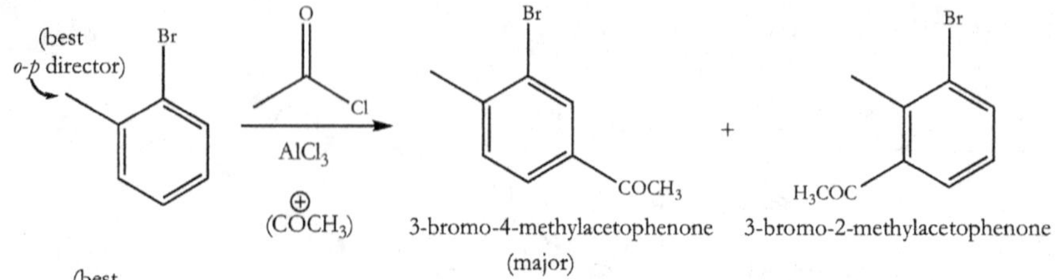

(best *o-p* director)

AlCl₃

$\overset{\oplus}{(COCH_3)}$

3-bromo-4-methylacetophenone
(major)

3-bromo-2-methylacetophenone

(b)

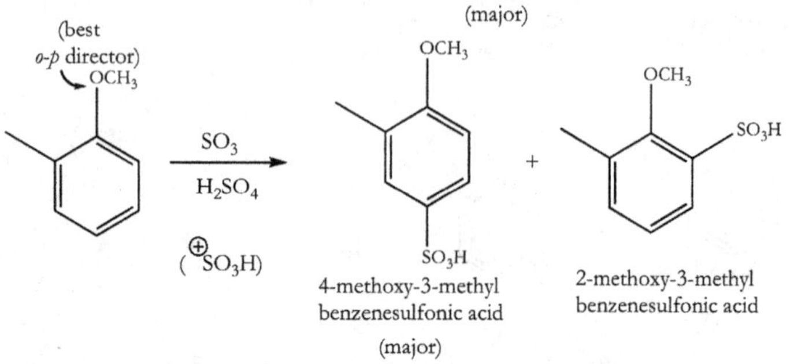

(best *o-p* director)
OCH₃

SO₃
H₂SO₄

$(\overset{\oplus}{SO_3H})$

4-methoxy-3-methyl
benzenesulfonic acid
(major)

2-methoxy-3-methyl
benzenesulfonic acid

(c)

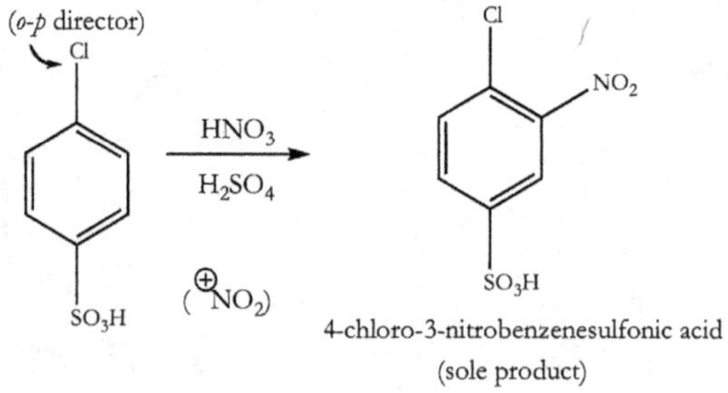

(*o-p* director)
Cl

HNO₃
H₂SO₄

$(\overset{\oplus}{NO_2})$

4-chloro-3-nitrobenzenesulfonic acid
(sole product)

Check: (1) Structures drawn correctly, (2) the most activating *o-p* directors and electrophilic species identified, (3) the most activating substituent selected as director for the incoming electrophile. (4) products proposed are consistent w/EAS theory.

Problem 7.14.

Given:

(a)

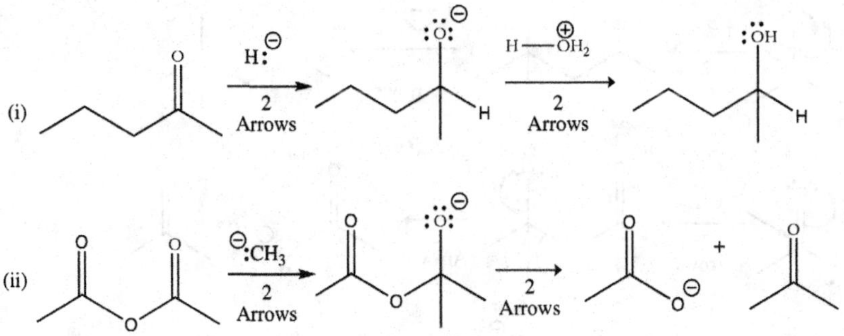

(i)

(ii)

(b)

(i)

draw intermediate   complete structure

(ii)

draw intermediate
I

draw intermediate   draw structure of
II   2-methyl-2-propanol

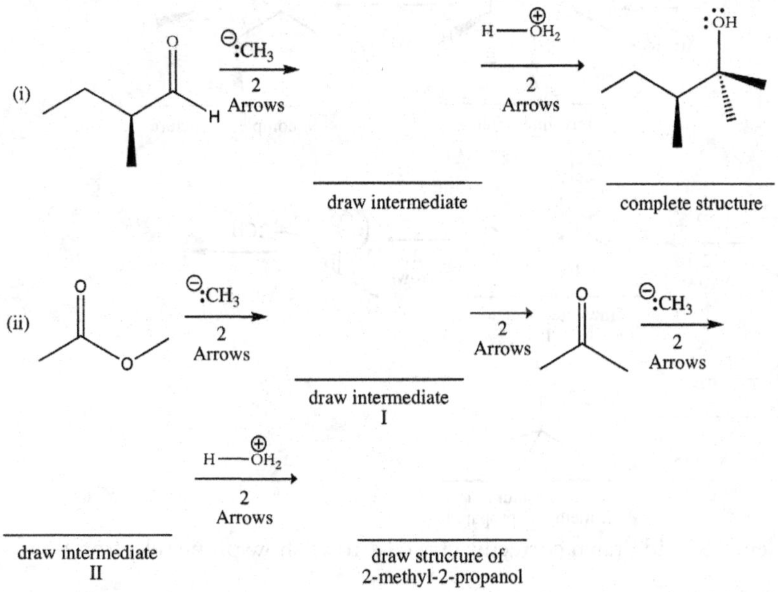

Find: (1) mechanism for (a) (i) and (ii), (2) structures of all species and mechanism for (b) (i) and (ii).
Plan: (1) Draw all species, including intermediates. (2) Identify nucleophile, electrophile, leaving group (if applicable). (3) Use curved arrow formalism to show movement of electron pairs enroute to product.

Solve:

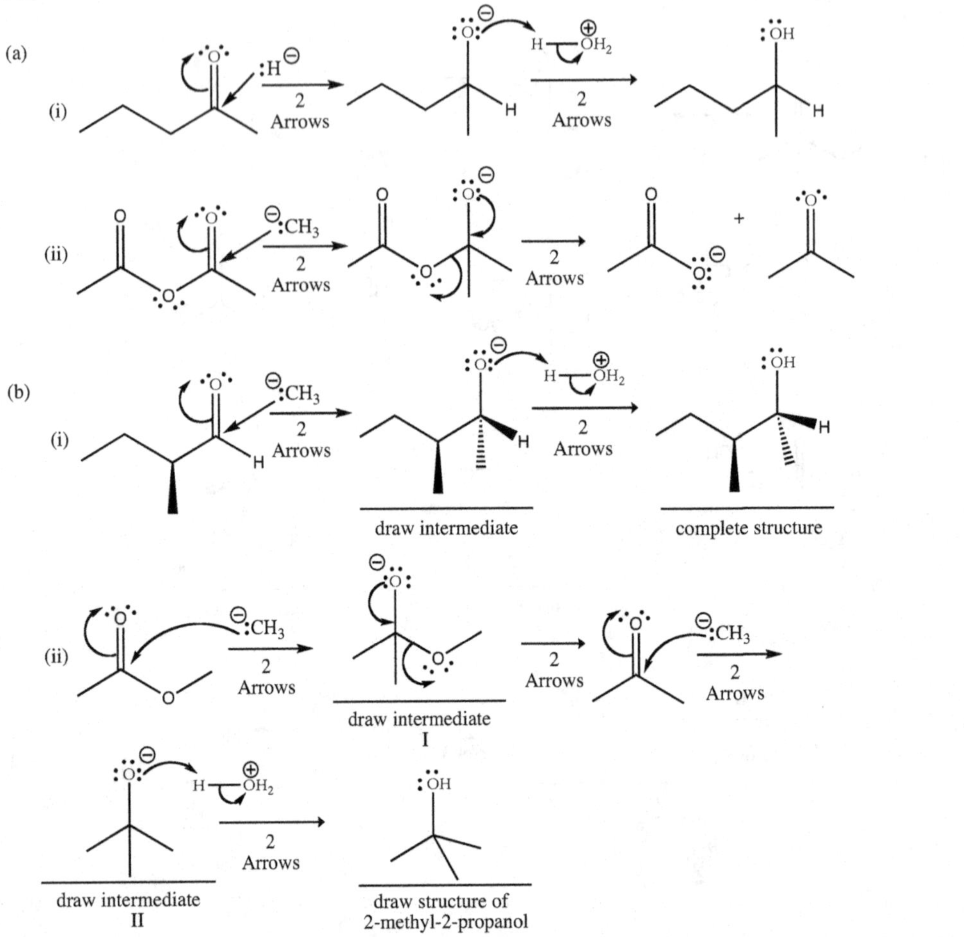

(a)

(i)

(ii)

(b)

(i)

draw intermediate   complete structure

(ii)

draw intermediate
I

draw intermediate
II

draw structure of
2-methyl-2-propanol

Check: All species identified and drawn correctly; curved arrows show proper electron flow.

Problem 7.15.
Given:

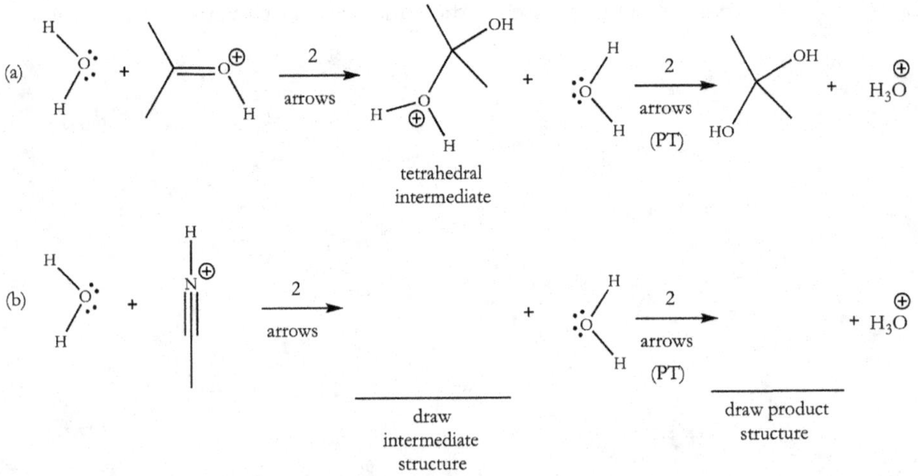

(a)

tetrahedral
intermediate

(b)

draw
intermediate
structure

draw product
structure

Find: (1) mechanism for (a), (2) structures of all species and mechanism for (b).
Plan: (1) Draw all species, including intermediates. (2) Identify nucleophile and electrophile. (3) Use curved arrow formalism to show movement of electron pairs enroute to product.

Solve:

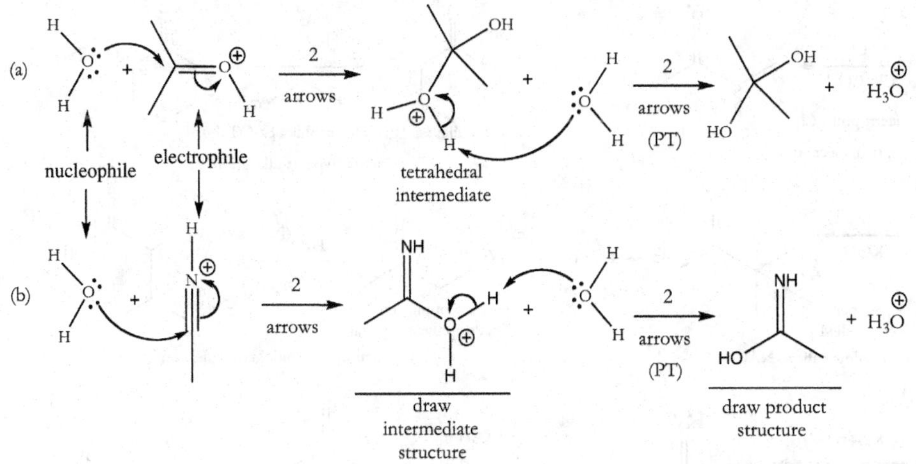

(a)
2
arrows
2
arrows
(PT)
nucleophile
electrophile
tetrahedral
intermediate
(b)
2
arrows
2
arrows
(PT)
draw
intermediate
structure
draw product
structure

Check: All species identified and drawn correctly; curved arrows show proper electron flow.

Problem 7.16.
Given:

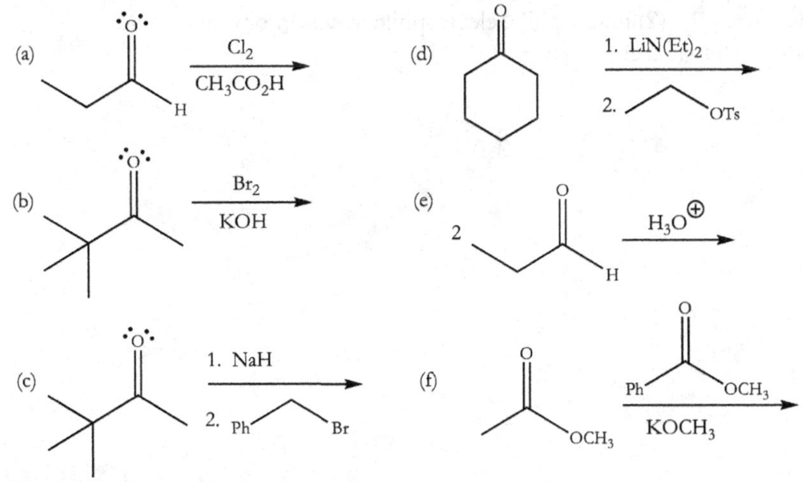

(a)
Cl2
CH3CO2H
(d)
1. LiN(Et)2
2.
OTs
(b)
Br2
KOH
(e)
2
H3O⊕
(c)
1. NaH
2. Ph
Br
(f)
Ph
OCH3
KOCH3

Find: Reaction products for (a-f).
Plan: (1) Draw all species. (2) Identify nucleophile, electrophile, reaction type and reagent function. (3) Based on knowledge of carbonyl reaction trends, predict the product(s).

Solve:

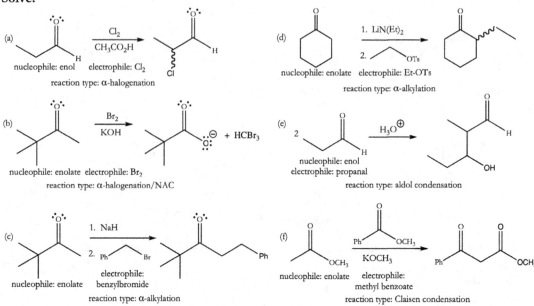

(a)
Cl_2
CH_3CO_2H
nucleophile: enol electrophile: Cl_2
reaction type: α-halogenation
(b)
Br_2
KOH
+ $HCBr_3$
nucleophile: enolate electrophile: Br_2
reaction type: α-halogenation/NAC
(c)
1. NaH
2. Ph Br
nucleophile: enolate
electrophile: benzylbromide
reaction type: α-alkylation
(d)
1. $LiN(Et)_2$
2. OTs
nucleophile: enolate electrophile: Et-OTs
reaction type: α-alkylation
(e)
2
$H_3O^{\oplus}$
nucleophile: enol
electrophile: propanal
OH
reaction type: aldol condensation
(f)
OCH_3
KOCH3
nucleophile: enolate
electrophile: methyl benzoate
reaction type: Claisen condensation

Check: (1) Structures drawn correctly, (2) nucleophile, electrophile, reaction type and reagent function identified, (3) products drawn accurately.

Appendix A – Reducing Agents

A.1. Reducing Agents

In organic chemistry, reagents which deliver electrons, H atoms, or H:⁻ ions are the most common reducing agents. Hydrogen atom delivery is usually accomplished via metal catalyzed hydrogenation, while reagents which deliver electrons or H:⁻ ions include metals such as zinc, aluminum and boron. Typical examples are found in Table A.1.

Table A.1. Reducing Agents.

Hydrogen Atom (H·) Donors		
Reagent/ Reagent Set		Reductive Function
Palladium	H_2, Pd on C	- reduce alkenes/alkynes to alkanes - reduce benzylic aldehydes and ketones to 1° and 2° alcohols
Platinum	H_2, Pt on C	- reduce alkenes/alkynes to alkanes - reduce benzylic aldehydes and ketones to 1° and 2° alcohols
Nickel	H_2, Ni, EtOH, 70 atm	- reduce imines to amines
Rhodium	H_2, Rh, 1000psi	- reduce aromatic rings to cycloalkanes
Lindlar	H_2, Pd on $CaCO_3$, quinoline, $Pb(OAc)_4$	- reduce alkynes to cis alkenes
Electron Donors		
Reagent/ Reagent Set		Reductive Function
Zinc (Clemmensen Reduction)	Zn, HCl	- reduce aldehydes and ketones to alkanes
Hydride (H·) Donors		
Reagent/ Reagent Set		Reductive Function
Sodium, Lithium (Dissolving Metal Reduction)	Na, NH_3 Li, NH_3	- reduce alkynes to trans alkenes
Hydrazine (Wolff-Kischner Reduction)	NH_2NH_2, OH⁻	- reduce aldehydes and ketones to alkanes
Boron	Li or $NaBH_4$ $NaBH_3CN$	- reduce aldehydes and ketones to 1° and 2° alcohols - reduce imines to amines
Aluminum	$LiAlH_4$	- reduce all carboxylic acids, esters, or aldehydes to 1° alcohols and ketones to 2° alcohols - reduce nitriles to 1° amines and amides to 1°, 2°, or 3° amines
	$LiAlH[CH_2CH(CH_3)_2]_2$ (DIBAL)	- reduce esters to aldehydes
	$LiAlH[OC(CH_3)_3]_3$	- reduce acid halides to aldehydes

Here are some problems to work:

Problem A.1 Identify the products likely formed when 2-butanone reacts with:
a. 1. LiAlH$_4$ 2. H$_3$O$^+$ b. 1. NaBH$_4$ 2. H$_3$O$^+$

Problem A.2 Identify the products likely formed when 2-butyne reacts with:
a. H$_2$, Pd/C   b. H$_2$, Lindlar catalyst   c. Li, NH$_3$

Solved Problems

A.1

Given: but-3-en-2-one, a. 1. LiAlH$_4$ 2. H$_3$O$^+$ b. 1. NaBH$_4$ 2. H$_3$O$^+$

Find: Identify the products likely formed when 2-butenone reacts with: a-b.

Plan: (1) Draw reactants, (2) identify reductive sites on substrate, (3) identify reagent function, (4) predict the products.

Solve:

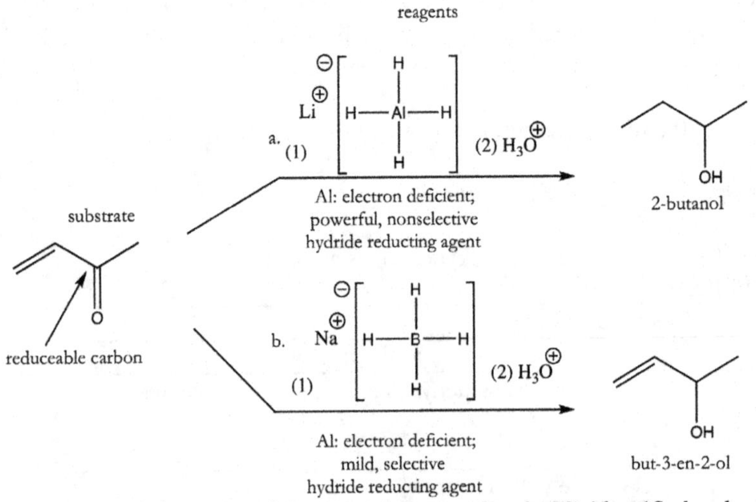

Check: (1) Substrate and reagents drawn correctly, (2) identified reductive sites on substrate, (3) identified reagent function, (4) predicted the products in accord w/reduction ability.

A.2.
Given: but-3-en-2-one, a. H₂, Pd/C   b. H₂, Lindlar catalyst   c. Li, NH₃
Find: Identify the products likely formed when 2-butyne reacts with: a-c.
Plan: (1) Draw reactants, (2) identify reductive sites on substrate, (3) identify reagent function, (4) predict the products.
Solve:

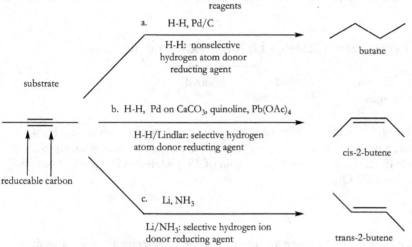

Check: (1) Substrate and reagents drawn correctly, (2) identified reductive sites on substrate, (3) identified reagent function, (4) predicted the products in accord w/reduction ability.

Appendix B – Oxidizing Agents

B.1. Oxidative Processes

Oxidation is defined as the net gain of oxygen atoms or loss of hydrogen atoms from an organic molecule while reduction is the net loss of oxygen atoms or gain of hydrogen atoms. Here is an example:

$$2MnO_4^- + CH_3(CH_2)_4CH=CH_2 \quad 2MnO_2 + \underbrace{CH_3(CH_2)_4COOH + CO_2}_{} \text{ (unbalanced)}$$

| oxidizing agent | reducing agent | reduced | oxidized |

In this reaction, Mn (in MnO_4^-) is reduced while the alkene is oxidized to a carboxylic acid and CO_2. In this case, permanganate, a powerful oxidizing agent, oxidizes the alkene (reducing agent) to a carboxylic acid and carbon dioxide (net loss of 3 H; net gain of 4 O) while Mn is reduced from an oxidation state with a net loss of 2 O.

B.2 Oxidizing Agents

As defined earlier, reagents that cause a loss of electrons in a substrate are termed oxidizers. In organic chemistry, many of these reagents provide oxygen to substrates which reduces the electron density at the site of introduction. If we can identify and draw the Kekule structures of these chemical species, we can then predict how they might react with a given substrate. Generally, there are two classes of oxidizing agents, inorganic and organic.

Although many reagents have been developed to oxidize specific substrates, chromium, osmium and manganese heavy metal based agents as well as silver oxide (Ag_2O), periodate (IO_4^-), hydrogen peroxide (H_2O_2), water and ozone (O_3) are some of the typical oxidizers commonly discussed in introductory organic chemistry. Alcohols and peroxyacids are also oxygen delivery reagents. These reagents are normally milder than the heavy metal oxides and do not typically overoxidize the target molecule. Let's examine a few of them to see how their structure influences their role as oxidizers. Some examples follow in Table B.1 with the scope of their oxidative tendencies.

Transition Metals

	Metal Oxide	Oxidative Function
Chromium: (acidic conditions)	CrO_3: [structure] , H_2CrO_4: HO—Cr—OH [structure] , $Na_2Cr_2O_7$: NaO—Cr—O—Cr—ONa [structure]	- oxidize 1° alcohols and aldehydes to carboxylic acids, 2° alcohols to ketones
	PCC: [pyridinium chlorochromate structure] ·CrO_3	- oxidize 1° alcohols to aldehydes
Osmium:	OsO_4: O=Os=O [structure]	- oxidize alkenes to cis-1,2-diols
Manganese:	$KMnO_4$: KO—Mn=O [structure] (acidic conditions w/ C=C bond cleavage) (basic conditions)	- oxidize alkenes (mono- and 1,2-disubstituted) to carboxylic acids - oxidize alkenes (trisubstituted) to a ketone and carboxylic acid - oxidize alkenes (tetrasubstituted) to ketones - oxidize alkenes to 1,2-diols
Silver:	Ag_2O: Ag—O—Ag	- oxidize aldehydes to carboxylic acids

Reagent Sets

	Oxygen Source	Oxidative Function
Hydroboration (peroxide)	1. BH_3, THF 2. HO-OH, OH-	- oxidize alkenes to alcohols
Oxymercuration (water)	1. $Hg(OAc)_2$, H_2O, THF 2. $NaBH_4$ (If ROH is used instead of H_2O)	- oxidize alkenes to alcohols - oxidize alkenes to ethers
ozone	[ozone structure] , Zn (w/C=C bond cleavage)	- oxidize alkenes (mono- and 1,2-disubstituted) to aldehydes - oxidize alkenes (tri- and tetrasubstituted) to ketones
periodate	$NaIO_4$: NaO—I=O [structure] (with C-C bond cleavage)	- oxidize 1,2-diols to aldehydes
hypochlorite	Na^+ O-Cl, CH_3CO_2H	- oxidize 2° alcohols to ketones
Peroxyacid m-chloro perbenzoic acid	RC(O)OOH [m-CPBA structure]	- oxidize ketones to esters - oxidize alkenes to oxiranes

The oxidizing agents shown in the table have one thing in common, <u>they all contain oxygen.</u> Let's apply what we know to solve some problems.

Problem B.1 Identify the products likely formed when 1-butanol reacts with:
a. PCC b. $Na_2Cr_2O_7$

Problem B.2 Identify the products likely formed when 2-butene reacts with:
a. $KMnO_4$ b. OsO_4 c. 1. $Hg(OAc)_2$, H_2O, THF 2. $NaBH_4$

Solved Problems

B.1
Given: 1-butanol, a. PCC b. $Na_2Cr_2O_7$
Find: Oxidation products of the reaction.
Plan: (1) Draw reactants, (2) identify oxidative/reductive sites on substrate, (3) identify reagent function, (4) predict the products.
Solve:

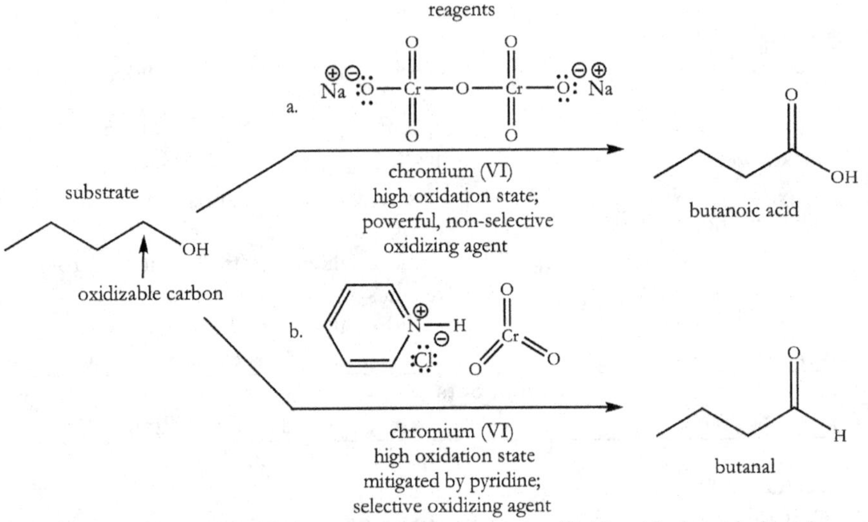

Check: (1) Substrate and reagents drawn correctly, (2) identified oxidative sites on substrate, (3) identified reagent function, (4) predicted the products in accord w/oxidation ability.

B.2

Given: 2-butene, a. $KMnO_4$ b. OsO_4 c. 1. $Hg(OAc)_2$, H_2O, THF 2. $NaBH_4$

Find: Identify the products likely formed when 2-butene reacts with: a-c.

Plan: (1) Draw reactants, (2) identify oxidative/reductive sites on substrate, (3) identify reagent function, (4) predict the products.

Solve:

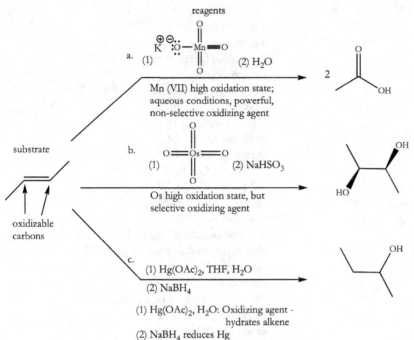

Check: (1) Substrate and reagents drawn correctly, (2) identified oxidative sites on substrate, (3) identified reagent function, (4) predicted the products in accord w/oxidation ability.